LINDSAY SCOTT CISNEY

CHRONICLES OF THE GODS

BOOK ONE

CHRONICLES OF THE GODS

BOOK ONE

LINDSAY SCOTT CISNEY

CITI OF BOOKS

CITIOFBOOKS, INC.
3736 Eubank NE Suite A1
Albuquerque, NM 87111-3579
www.citiofbooks.com
Hotline: 1 (877) 389-2759
Fax: 1 (505) 930-7244

Ordering Information:

Quantity sales. Special discounts are available on quantity purchases by corporations, associations, and others. For details, contact the publisher at the address above.

Printed in the United States of America.

ISBN-13: Paperback 979-8-90124-290-2
eBook 979-8-90124-291-9

Library of Congress Control Number: 2026921366

CONTENTS

Part 1: The Journey to Terra Nova in the Cervantes Star System

Part 2: The New World

Part 3: Lost in the Past and in Space-Time

Part 4: Becoming The Mother

It was always our mistake, perhaps our biggest one, to assume that the realms of knowledge would forever be separated from things we considered "unknowable."

L. Scott Cisney
Pleasant View, Utah
April 2025

PART ONE

The Journey to Terra Nova in the Cervantes Star System

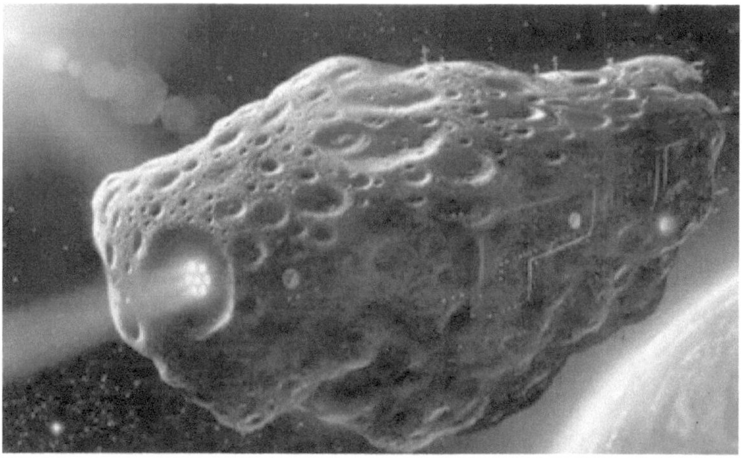

PROLOGUE

Eden, the Earth, 10,912 BCE (Before the Current Era)

THE SENIOR ASHTEE, or Lords, were slowly gathering at the appointed time in the main conference room of the great hall of the governance building. Elotim waited patiently, sitting in the front of the large room.

"Everyone, if you could please find a seat. We have a very serious issue to discuss, and we will undoubtedly have some lengthy discussions," Elotim said, with no humor in his voice.

After a few more minutes, the last of the Lords found their seat.

Mareye stood, dressed in her ceremonial robes, and began.

"I believe we all know about the news from our space-based colleagues. About two local weeks ago, they detected an unusual fast-moving three-headed comet that entered this solar system. The comet's orbital period was shifted by the gravitational pull of the largest gas giant. It is now on an intercept with this world, approximately twenty-five local weeks hence. At the moment, because of the instability of its orbital trajectory, we are not certain where these bodies will impact. If they impact the ice sheets covering the northern latitudes of the westernmost continent, it will melt these sheets instantly, causing sea levels to rise by two hundred and fifty cubits within a few local rotations. Our location here will be inundated, as well as our satellite outposts around the globe. We are, as you are all aware, not completed with our work here. The final addition to the genome of the preeminent species has not been finished. As a minimum, this step must be completed. The remaining steps can be accomplished by our junior colleagues—this one cannot. If we leave Eve here with her new husband, she will need to lead them southward. We recommend that she return to the original home. As you know, it is more than a three-hundred-thousand-cubit journey southward back to the original home of the upright peoples in the Great Cracked Valley. This will be a great journey for her people, and we will need to prepare them appropriately. We believe this is the most prudent step. The destruction of all our facilities by the ensuing deluge will completely halt all our remaining activities. Additionally, the comets will almost certainly break apart due to the gravitational tidal forces of this world and its moon. It will be unsafe for us to remain either on the surface or in orbit for a long duration. Eve, do you agree to remain behind and lead this species through the formation of their first empires?"

"Mother, I agree to submit to your wishes and shall remain and guide this species through the crucial first steps. I know The Ones will be with me and these people, my people," Eve of Eden answered as she bowed her head to her mother, leader of what the local people referred to as the Sky Lords.

CHAPTER 1

Farewell to Earth

Port Vandenberg, California, July 1, 2340, CE (Current Era)

THE GLARE OFF the Pacific was blinding. As Sam and his family waited at the bus stop to the *John Glenn*, he had another fellow traveler take a final photograph of him with his wife, family, and the FBS (Free Bloc Ship) *John Glenn* in the background. This would be the last photograph his family and their descendants would have of their family on Earth.

The dying Earth had suffered so much. No one debated that anymore. There were no hints of doubt from even the strongest deniers. The "Last Days" were truly here, and destruction and extinction awaited humankind. To prevent this, and save the species, humankind built the largest spaceship ever, the FBS *Calypso*, designed to travel to a planet outside the solar system. In fact, the *Calypso* was not really a spaceship; it was a five-kilometer diameter asteroid, waiting just inside the asteroid belt. By building living quarters inside the asteroid, the occupants were protected from deadly cosmic radiation, which they otherwise would have been exposed to for centuries.

Sam, his wife Riley, and their daughter Terra, shortened from "Terra Nova," were among the three thousand hand-selected humans boarding transports for the *Calypso*. The *John Glenn*, their lift vehicle, would take them to O'Rourke Station, the spaceport at the Lagrange Point between the Earth and the moon. From there, it was a 191-day trip on one of the mighty "sailing ships" to the *Calypso*. The sailing ships used solar sails to accelerate to more distant points in the solar system. The large solar sails captured charged ions streaming off the sun which imparted them with a small amount of momentum. Over time, this allowed these types of ships to achieve impressive delta v's.

The transport bus dropped over a hundred people off at the ramp to the elevator, which would take them to the passenger entrance of the *John Glenn*.

The travelers were talking excitedly with one another as they walked up the ramp. The loudspeaker repeated directions to the families to continue to the elevator as a group.

Terra was looking around her at all the other families and at the giant rocket in front of them, taking it all in. She was thrilled about the adventure ahead.

"Daddy, will our new home be as beautiful as this one?" she asked.

Sam knew there was no way to answer that question. Their destination had been picked to give them the best chance for survival, but no one knew the actual condition of their new home.

He said, "Yes, Terra. In time, I am sure it will be a beautiful place."

"I want a dog when we get there."

Sam and Riley looked at their daughter and smiled. They knew that for Terra, this was just a great adventure. For Sam and Riley, leaving was a melancholy experience. They were leaving all their family members and friends. Within a few hundred years, extinction awaited humankind on the Earth.

Sam stood on the loading ramp before entering the *John Glenn*. He turned around to get a last view of the wide-open spaces of land and sky. He would never see this place again. His life would now be steel domes and hallways. The melancholy burst out into the open as a sudden wave of sadness hit him, and he started crying uncontrollably. He recalled a line from the New Testament lamenting the world's condition, exclaiming, "Alas Babylon!" "Babylon," the Earth, was being gathered up for destruction. As tears coursed down his cheeks unchecked, he furiously wiped them in embarrassment; but when he looked around, he noticed other people doing the same. Like his immigrant ancestors from the British Isles and Germany, hundreds of years earlier, Sam and all the others were leaving everything they ever had known forever.

Given the accelerating pace of the destruction of the Earth's ecosystem, the fate of their friends and families left behind was uncertain, though most experts believed the Earth would not survive for more than three or four generations.

CHAPTER 2

Parting Is So Painful

Aurora, Colorado
A Few Days Earlier

"WE'RE GOING TO be late, honey!" Riley said, exasperated at Sam's chronic lateness.

Sam sighed. "I'm just trying to get Terra to wear the new dress that your mother bought her. She doesn't want to, but I finally got her to agree. I'm sorry. I don't think we're going to be late."

As they bundled into the robocar, they were all finally calming down. This was going to be a tricky evening anyway.

Emotional *was probably a better word,* Sam thought.

When they reached his in-law's house on the outskirts of Aurora, the sun was setting over the distant mountains, sending long shadows over the prairie to the east.

It always looks so "alive" at this time of day, Riley thought.

She smiled as she thought of childhood memories of June when school was out, and she and Anne would play until after the sun went down on those long, early summer days.

Anne. How do I make this right with Anne? Riley thought.

Anne was decidedly not on board with leaving and never returning. They had hardly been able to talk to each other the last few months as the departure day approached. Riley desperately wanted this evening to be one of love, reconciliation, and sweet parting.

How can I make it through this evening? How do you say goodbye when goodbye is forever? Riley thought.

"Sam, Riley, Terra, how are you doing?" Sam's father-in-law, Edward, asked.

"We're good, Ed," Sam said. "Listen, I hope everything goes well this evening. I know this is hard on everyone."

"It's Kelly and Anne I am most worried about, but you already know that. Let's try to keep everybody focused on everything but, you know, leaving," Edward said.

Sam and Edward exchanged a pained smile at the last word.

Leaving—yes, that was the word everyone was thinking about but couldn't say out loud, Sam thought.

As they settled into their seats for dinner, Anne sat stiffly at one end, as far as she could from her sister Riley.

"I'm not all that religious as you all know, but perhaps we could say grace as they used to say in my grandmother's day," Edward said as they all became silent, waiting for him to say something that indicated it was acceptable to begin eating.

“Oh, that is so sweet, Daddy,” Riley said with a sniffle and a tear.

Edward picked up his wife’s hand and then Anne’s and motioned for them all to join hands.

“Our Heavenly Father, we join together to eat this lovely meal that we have been given. We do this as a family, at a time in our lives of great stress and uncertainty, but with love in our hearts for one another. Please sustain us tonight and in the days, weeks, and years to come when we will be apart. Father, our hearts are heavy, please sustain us, leave us with only good memories of love and happy times we shared.” Edward paused as Anne broke down into sobbing. “We ask for thy blessings to be on this food, for which we thank thee. Amen,” Edward finished as everyone excused themselves to reach for tissues on the kitchen island.

“Thank you, dear,” Kelly said to her husband, who also grabbed a tissue.

As dinner proceeded, the mood lightened as each shared family memories at the urging of Edward.

At almost nine o’clock, Sam said, “I guess . . .” He paused as his emotions again got the best of him. “I guess we should get going. We have a big day tomorrow. We have an early flight to California, and we still have a few things to pack.”

As they gathered in the doorway, Anne came up to her sister and hugged her; then she grabbed her niece and hugged her.

"Be good, little girl, who is not so little anymore," she said to Terra who had now begun crying again.

"I will, Auntie Anne, I will. Take care of Winnie for us," Terra said as her dog's ears perked up upon hearing her name.

As they rode into the darkness in the robocar back to their home in Denver, they found it impossible to talk to one another through the tears and sobbing.

Forever, *what was that word supposed to mean for human beings? We can hardly think of anything beyond next week,* Sam thought.

CHAPTER 3

Liftoff!

"FIVE, FOUR, THREE—IGNITION sequence starting—two, one, liftoff! We have liftoff at 09:21 Pacific Standard Time," the voice on the intercom said.

The roar and vibration of the powerful liquid engines extinguished all possibility of a conversation. Terra began crying in fear. Her mother reached over and grabbed her daughter's hand and squeezed it. Sixty seconds later, the passengers were exposed to a violent jolt as the intercom announced, "Stage one cutout and separation achieved. Stage two ignition successful!"

By now, the travelers were getting used to the incredible sounds and acceleration. In another six minutes, all the rumbling stopped as they achieved Earth's orbit.

The large screen at the front of their section showed the third stage dropping away with the Earth looming large below it.

"Mommy, I see France! I'll bet I can see Paris if I squint! Mommy, you remember Paris, all the cool buildings and monuments? I am so glad we got to go there for ship training. I'm going to miss Elyse. She was so nice," Terra said.

"Yes, dear, I remember. It would have been nice to see more of Europe, but at least we saw France," Riley answered.

"Trans O'Rourke insertion burn in three, two, one!" The now-familiar intercom voice announced as they all felt another jolt of acceleration.

We are on our way now. This is now real. There is no going home, Sam thought as he again was overcome with emotion.

CHAPTER 4

The World of the Twenty-Fourth Century

THE JOURNEY TO O'Rourke Station at one of the Lagrange points between the moon and Earth spanned two days. Sam found himself bored while his wife and daughter spent most of their time looking out the large bay windows in the Inquisition Victims Memorial Lounge.

"Mommy, I can see Colorado!" Terra exclaimed.

Riley looked out the window and for a moment was overcome. Her parents were down there somewhere, as well as Sam's. Parting with them was the hardest thing she had ever done in her life. Her sister, Anne, was angry with her for leaving, though there had been something of a reconciliation at their farewell dinner at her parents' house in the Denver suburbs.

Riley thought, *I am not sure anyone could understand how difficult it was to leave and never return. It's like dying. Everyone dies and it is one of the hardest things we experience in life. But this feels harder. It was like choosing to die at an early age, and that made it impossible to understand. And Anne did not understand.*

She had tried to explain her reasons for volunteering for the voyage, but in the end, all that was apparent was she was leaving and would never return. Her parents seemed twenty years older than the last time she saw them. The prospect of never seeing their daughter again had taken a heavy toll on them.

"Mommy, why are those places just brown?" Terra asked.

Riley pointed at these areas and said, "The brown areas were once green and living, but not anymore. They were once called Russia and China. The second nuclear war between the Russians and Chinese, who started the first nuclear war against the Free Bloc, caused these areas to die. Scientists estimate that it might be safe to go into those areas again in twenty-nine thousand years. The real damage, however, was caused when the attacks destroyed nuclear power facilities, which put thousands of tons of highly radioactive isotopes into the atmosphere and poisoned the land upon which it fell. As a result of climate change and these wars, the Earth is dying."

"We'll make sure that never happens on our new Earth, Mommy," Terra solemnly proclaimed.

"That's right, honey. We must do things differently this time," Riley answered, with tears welling in her eyes.

Sam, meanwhile, had gone to the bar and was sipping a rum and Coke. He had finally figured out that the way to kill his boredom was to make a friend. *After all*, he thought, *I am going to spend the rest of my life with these people.*

"So your great-great-grandfather was one of the negotiators that brokered the unconditional surrender of the Zealots that ended the second civil war?" Sam asked.

"Yup. To hear him talk about the attitude of those people, their arrogance, even in utter defeat, was just so unbelievable," Teddy, his drinking partner slurred. "I mean they still wanted their tax-exempt status for their political lobbying organizations! That was part of their attempts to overthrow the old US Constitution—pretend they were not making political commercials, tax-free, while their competition was hobbled by tax constraints."

"The old US Constitution was deeply flawed. The best thing that ever happened to our country was the new constitution."

"Oh, I agree. The Senate, the Electoral College, legalized bribery of elected office holders—" Teddy started.

"The lifetime appointments of clearly unqualified and biased jurists to the Supreme Court," Sam finished.

Both fell silent as the liquor tempered the anger they both felt rising inside of them as they thought of the deaths of four million people, mostly LBGTQ people, atheists, agnostics, feminists, opposition political leaders, Muslims, and fringe religious cults, which resulted from the inquisition of the 2030s, all justified in the name of national security by hideous rulings from the US Supreme Court. This outrage led to the second US Civil War of the 2040s, which caused another five million deaths. Sam had a minor in Middle Times US History, so he had a grasp on the problems in the US in the twenty-first century.

"Here's to getting it right from the outset, this time around on our new world!" Teddy raised his glass, and Sam did the same.

Upon their arrival at the giant O'Rourke Space Station, they were transferred to temporary quarters while they waited for their planetary transport ship, an interplanetary "sailing vessel" that employed solar sails along with booster thrusters. The sailing ships used booster rockets as they began accelerating and then again as they decelerated when they drew near to their destination.

A week to the day later, they boarded the *Yankee Clipper* and departed the space station bound for their starship, the FBS *Calypso*.

CHAPTER 5

The Multigenerational Starship FBS *Calypso*

AS SAM, RILEY, and Terra drew near the end of their journey to the *Calypso*, located just inside the asteroid belt, the excitement became palpable. They knew this was going to be their new home. They had been given extensive training on the ship in trainers located at the European Space Agency in Paris.

NASA and the European Space Agency (ESA) worked together on the *Calypso* and on all other aspects of the mission, including crew selection and training. The ESA had built a mock-up of the *Calypso* living and working spaces to aid in crew training. This mock-up was called a trainer. It included the engineering, reactor, environmental, communication, and living spaces. Every person who was selected for the crew was given real-life training on every relevant system. The training took a full year for adults and was also required for children over twelve. Younger children were given abbreviated training appropriate for their age.

Despite the training, when the massive *Calypso* came into view, it took everyone's breath away.

Sam looked at various antennae equipment on the side they were approaching. They were going to enter a giant hangar through doors that were now wide open. This allowed shuttles and transport vessels like the *Yankee Clipper* to enter and leave the ship.

The communication antennae on this side of the ship were duplicated on each side and both ends of the ship. This was of special interest to Sam as his responsibility was communication. He was the lead communication engineer.

Sam attempted to look at the aft end where the giant nozzles would be located, but they were just too close to the ship for him to see very far aft. He strained his eyes to see the front end, but again, the asteroid was so large and the front end was so distant that he could see none of the equipment he knew would be located there. Everything beyond a kilometer or so was just too far away to make out many details. No individual features on that end were even visible.

Sam noticed it was not spherical, but instead was potato shaped. The asteroid had been modified so that it became more symmetrical, thus making the line of action of the thrust align with the long axis of the asteroid. On the sides, front, and back end were batteries of telescopes and radar arrays that gave a complete view of space around the ship.

Terra was amazed at the size. “Mommy, it is so big! How will it ever get moving?”

"It will get moving by continuous thrust from the ion engines. We will move slowly at first, then we will gain more and more speed. Also, as we leave, we will get an assist from four giant solid rocket motors that will help us to get moving. When they burn out, they will be cut loose. We have several sets of these rockets to help us at critical times when we want to slow down or speed up."

"Mommy, will we go to sleep when we get on board?"

"No, honey, we will be in the first crew, so we will be awake until you are eighteen years old. They want people to be at least eighteen years old before they are put into hibernation," Riley explained.

Terra looked relieved then asked, "So there will be other children for me to make friends with?"

"Yes, Terra. You met some of them when we went to Paris for crew training. Do you remember Hatcher and Cassia?" Sam asked.

"Yeah, a little bit. It was a long time ago, Daddy."

"Last summer was not really a long time ago, sweetie," Sam said, smiling at Terra.

"It is for me. Summer lasted forever and then there was school, so it was a long time!"

"Well, you will see them soon. I think they got here a week ago, if I remember the boarding manifest correctly," Sam said.

The orientation was held in the shuttle bay the day after they arrived. The first "awake captain," Captain

Sandra Simmons, started the orientation by greeting the arrivals for that week.

"I would like to welcome you all aboard the *Calypso*. You are the third group of twenty-one groups that will be boarding the *Calypso* over the next four months. The first six groups are families with children. We wanted it this way so that the children and their families could have some time to get to know one another better before the bulk of the crew arrived. Families are going to be the backbone of our life aboard ship and once we get to the settlement planet. I will do everything I can to help each of you get comfortable with the ship and your duties. Everyone, even children, will have duties to be performed on board. Each of you children will have much expected of you. You will go to school, but you will also be learning about your careers and your duties on the ship when you become old enough to carry them out. Think of pioneer children crossing the great plains of North America in the nineteenth century. Those children had much expected of them also. The same will be true of you children and all the crew of this ship. We are making an extraordinarily long voyage to an unknown world. Once we get there, there will be even more expected of you. Having said that, I still want you to make friends and have some fun. We will need to buoy each other up, for there will be many moments when we will get homesick and long for the good Earth that we have all left. The hopes of humanity rest with each of you. I know you will meet the challenges successfully.

Right now, I want to turn the time over to my first officer, Peter Chan, our chief engineer. Peter, take it away!"

"Good morning, everyone. I am Peter Chan, the chief engineer. I have been involved in the design and fabrication of this vessel for more than fifteen years. Actually, it is closer to twenty years now. We have a good ship, the most advanced one ever built. It is a generational ship, meaning that many generations will call it home. Of course, the limitations discovered by Albert Einstein have dictated this—we cannot travel faster than the speed of light. Because of this, it will take us over a thousand Earth years to reach our destination."

Peter continued, "We will be using an ion engine that will continuously accelerate us to a maximum of 4 percent of light speed, unlike the old days of space travel where a ship was accelerated by solid or liquid fuel, which burned out quickly, making the vessels coast all the way to their destination instead of accelerating most of the way there. Our propulsion nuclear reactor will provide an almost limitless supply of xenon fuel. We take these xenon atoms and strip an electron-making xenon ions for our ion engines."

"Our ship was designed to have an expected life of over two thousand years. With proper maintenance, it will last much longer than that. We will need to decelerate for part of our journey so that we do not go whizzing past our destination! The deceleration phase will be aided by solar sails that will work to slow us down, along with our ion drive, which will also slow us down after we turn our ship around. This acceleration and deceleration will give us something else that we need—gravity. Gravity will be mimicked by the constant acceleration and deceleration of the ship. Our bodies need gravity for us to survive and thrive once we reach our destination, a world that will have approximately 97 percent of the Earth's gravity. It is anticipated, based on the hibernation schedule, that most of us will survive to see our new world. For me, I am going to enjoy a few years of good sleep! I have spent quite a few years under significant stress designing and building this ship, which ruined my sleep schedule. I do not know about all of you, but I am going to need the sleep!" Peter's last comment brought a smattering of laughter from the assembled crew.

The orientation went on for another two hours, at the end of which it was lunchtime. After lunch, Terra quickly left to find Cassia and Hatcher.

CHAPTER 6

A Ten-Year-Old on a Starship

TERRA WAS HAVING the time of her life. For a ten-year-old girl, the inside of the *Calypso* was a labyrinth of hallways, hatches, and equipment rooms, perfect for hide-and-seek. She had been on the ship for only ten days and had already reactivated her friendship with the two children she had met during crew training in Paris: Hatcher, a boy, and Cassia, a girl.

They played hide-and-seek for hours while their parents settled into their new jobs on board. School was still out, given that it was not yet January, the new school semester. On the ship, there would no longer be any real seasons, let alone days and nights, at least not like what they had on Earth.

Instead, the crew that was not in hibernation was divided into three groups—the A shift, B shift, and C shift. The ship required constant attention, despite its highly automated systems; hence the need for 24-7 coverage by the engineers and other staff. Each shift adjusted to a different twenty-four-hour "day." Hatcher and Cassia's families worked the same shift as Terra's family, the A shift.

The children took a break and sat down in the crew cafeteria.

"Terra, do you miss Winnie?" Cassia asked. Winnie was Terra's goldendoodle that her family had left behind.

"I do. I wish she could have come along. I wish she could have been chosen to be part of the domestic animal stock," Terra answered sadly.

"I had hoped my mom would pick Winnie for the animal stock," Cassia said ruefully.

Cassia's botanist mother was also responsible for the stock of animals the ship had taken on.

"They wanted working dogs. At least they have some cats, though. Cats are good emotional support animals," Cassia said with a smile. "One of the momma cats had kittens. We should go look at them."

Hatcher piped up, "I saw them yesterday. One of the baby cats is a calico. She has sweet colors, orange, black, and white! I wonder if the new world we are going to already has cats and dogs. It probably has dinosaurs! That would be cool. I want to have a pet dinosaur!"

"We would need to know they can be domesticated," Cassia said. Cassia was picking up knowledge from her mom.

"Does that mean they can be trained to be nice, like dogs?" Hatcher asked.

"Yes. Also, they must have a useful purpose. Cows give us milk. Chickens give us eggs. That is why they were domesticated a long time ago," Cassia said.

Terra looked at both. "Well, I just like animals, especially baby ones! Let's see the kittens!"

The three children made their way to the livestock area. They were looking at the tiny kittens when their ship comm devices let them know it was time for dinner. Schedules and discipline in performing assigned tasks were of paramount importance on their ship during its long voyage to their new home. This was constantly drummed into the children. Terra, Cassia, and Hatcher reluctantly bade farewell to each other and went their way through the hallways of the ship back to their parents' suites.

CHAPTER 7

The Message

AFTER DINNER, TERRA was required to shadow her father, who was her mentor, as he performed his shipboard duties. For most children, this meant becoming familiar with the jobs performed by one of their parents. It was believed that apprentices could learn the fastest by learning from their parents. Even at her young age, Terra was expected to learn complex duties. In this way, their technical education became relevant as they saw the importance of the jobs they would perform as adults and how it contributed to the smooth operation of the ship.

Sam and Terra had just arrived and sat down in the communications center or ComCent to begin the evening portion of their daily shift. This was the final three hours of their eight-hour shift. The "workday" consisted of work, breaks, meals, and more work. Another communication engineer, and her child, worked during Sam and Terra's breaks and meals. The breaking down of the workday into "chunks" was meant to prevent boredom and the resulting inattentiveness that accompanied less than fully engaged work. The lead communication engineer's job was to monitor all ship-to-Earth communications as well as communications between the Mars and Europa colonies. If a part of the system malfunctioned, they were responsible for fixing it, if they could, or for calling for help from communication techs if necessary.

The communications were either regular "light speed" voice and video streams received via radio and/or TV signals or through entangled particle communication (EPC).

EPC allowed for instantaneous communications without respect to the distance between the sender and receiver. The theory had been proposed in the twenty-first century but had only been developed for space travel in the last thirty years. Pairs of "entangled" particles were created in the old CERN particle collider through high-speed collisions of appropriate particles. The particles produced were identical but opposite in their spin and charge. It was discovered that a change in the spin of one of the paired particles would result in the opposite change in the other paired particle instantaneously, regardless of the distance between the particles. This odd feature of quantum mechanics allowed for a faster-than-light method of communication that greatly aided long-distance space travel.

The control panel in front of Sam began blinking on the light speed (conventional communication) data line. Given that they were still within reasonable range of Earth, Mars, and Europa, Sam thought it was a message for one of the engineering groups on the ship, probably answers to questions on system operations.

He put in the earpiece and pressed PLAY to hear the message, realizing that it was not any personal communication since it did not have a message header indicating such. As the message began to play, he was surprised that it was a series of tones of different amplitudes and frequencies. He decided to run the message through the message analysis program to determine its origin since neither the sender nor their location was specified, and to see if the message could be decoded into speech. As the output of the analysis scrolled down his screen, Sam stopped and stared.

The message originated from *outside* of the solar system.

The point of origin was estimated to be approximately fifty light-years distant, near the star HD 160691, known as Mu Arae or Cervantes.

"What the hell?" Sam said, startling his daughter who had been reading in a chair next to him.

"What's wrong, Daddy?"

Her dad ignored her and instead tapped an alert on his ship communicator, sending a message to the current awake ship captain, Sandra Simmons. Sam sent Terra back to their quarters as he waited for the captain and the other officers.

In less than ten minutes, Captain Simmons, Chief Science Officer Stuart Carroll, and Chief Engineer Peter Chan huddled around the communication console listening to the message. It was not understandable as speech or images, but it was clear that it was an organized set of tones, certainly not from any natural source. The message was beautiful melodic music, but unlike any earthly music. It was strangely hypnotic and drew all of them into the sounds. It took effort to break away from listening to the message. They all agreed that they experienced the same effects.

"Send the entire message back to the FB Space Agency and the FB Secure Communication Agency," Captain Simmons told Sam.

"Wouldn't they already have the message?" Peter asked.

"No," said Sam. "This was directed to us in a tight and very powerful signal."

"Wait," Stuart stammered, "how would they know our location and velocity with such accuracy to direct it to just us?"

"A good question," Captain Simmons said. "Let's let the experts on Earth try to decrypt this. "For now, no one else among the crew is to be told about this."

And so without fanfare, and after centuries of listening, humankind received its first extrasolar message.

CHAPTER 8

A Choice

IT WAS TWO weeks until the ship heard back from Earth concerning the meaning of the message, as it was now being referred to by the four individuals who knew about it. Sam was on his shift again, this time in the morning portion. Terra had started school with the other kids. Now, the only time she would spend with her father helping him in ComCent would be when school was in recess for one of the several holidays the ship observed.

On this occasion, a message appeared in the EPC tubes and was displayed on the main console screen. Again, Sam tapped an alert into his ship communicator, directed at the same three people as before. The captain responded that they would meet in the communications conference room to discuss the findings from Earth.

Sam had not yet opened the message, so all four were listening to it for the first time. As they sat down in the conference room, Sam queued the message up and played it in holographic mode. One of the people who appeared in the video was well-known; one was not. The president of the Free Bloc, who was immediately recognized, Simon Perrin, introduced Dr. Phillip Rogers, his science adviser.

Dr. Rogers began. "First, let me say that this message has been classified as top secret presidential access. No one outside of the four individuals addressed in the message header is cleared to know the contents of this message. The message you received has been decoded. The tones and amplitudes of the music had the meat of the message embedded in it. The music was intended to entice the listener to decode the real message. The arrival of this left us with quite a choice to make—stay the course and proceed to your original destination or change course and go toward the origin of the message. We have decided to change your destination to the first planet of the Cervantes star system, the home planet of the aliens who sent this to us. This is approximately fifty light-years distant, meaning a mission duration of 1,250 years at 4 percent of the speed of light. This is within the original mission specification, representing only a 10 percent increase in the original mission time. You should understand that the mission goals are the same—to find a habitable planet and establish a human colony to perpetuate the human race. The message senders indicated that they have a habitable world that would be suitable for our people. They have offered the second planet to us. They knew about your original mission and destination and invited us to visit their world, the first planet in their system, and to settle the second planet of their star system.

"Since we do not share entangled pairs of particles with them, any future communication will be via light speed communication, that is, with fifty-year message lag times and one hundred years before a return message could arrive back to the sender. The message they sent us included pictures of their planet and their people. We have attached the images to a file with this transmission. The images are perplexing. They look like us with only a few differences as far as we can tell. We do not understand why this is the case. We believe this race of people has sent probes to our planet and have been watching us for an extended period," he stated without further explanation.

"We do not know the reason they chose to contact us now. For your mission, however, this means very little has changed. Do NOT let the rest of the crew know about the message. Tell the crew that the destination change is due to updated information on the habitability of the planets in the Cervantes system. We anticipate that at some point we may have additional information from them and may or may not choose to share it with you. We sent a reply informing them to send messages to Earth from now on. If you receive any additional messages, do not open them. Instead, forward the messages to this office via quantum encryption. For now, carry on as before. Rogers out."

Captain Simmons, Stuart, Peter, and Sam sat in silence as the message ended.

"It looks like there is a file with the music, which we have heard, one with the message, which Dr. Rogers summarized, and there are images. Let's open the image files," Peter Chan said.

"OK," Captain Simmons answered.

The attached image file contained seventy-seven images. The very first image opened very slowly as it was a detailed holographic file. The holograph slowly began to form in the center of the conference room. As it grew clearer, it became apparent that the images were "people," *people who looked human.* Sam felt the hair on the back of his neck stand up as he stared at the image. There were two people in the image, both female, both smiling and stunningly beautiful.

The four of them stared for over a minute at the first holoimage. Oddly, there was music associated with it, which played in the background. It took significant effort to pull themselves away from the image. As they did so, they closed the first file and opened the second. For the next three hours, they opened each image file and felt a similar tug of emotions that made it difficult to continue.

Eventually, they finished and decided that they should get back to their other duties. As Sam left, he had a strange and powerful desire to share all of this with his daughter. He realized this would violate what the captain and Earth authorities had told them. He dismissed the thought but found himself coming back to it repeatedly over the next few days. He could not help but think that his urge to share the message with his daughter was being imposed on him from "outside."

"How crazy does that sound?" he muttered under his breath. But the thought kept pushing back into his brain.

Why do I feel this way? he thought.

CHAPTER 9

Hope for Our Children: Ten Years Earlier

"ONE MORE BIG push!" the lead nurse said, trying to sound encouraging.

"Ouch, I can't!" Riley screamed.

Dr. Tanner, with a slightly diminished bedside manner, curtly shouted, "Yes, you can, and you must!"

Riley pushed with everything she had, sure that she would pass out from the pain and strain of childbirth.

Why did I opt for natural gestation and childbirth? Riley thought, surprised she had the concentration to think given the pain.

And with this push, a baby girl entered the world. A world that had for three centuries teetered on the edge of doom.

Sam found himself relieved.

She's here now. The child they had debated over whether they should even have. *So many people nowadays are not opting to have children. Why had we done this?* Sam momentarily thought; then he looked at his daughter.

These thoughts left him as he and Riley looked at their perfect daughter. The tiny girl was red and purple and, well ... slimy!

But she is ours, and she is beautiful! he thought to himself.

Riley asked, "What should we name her, Sam? I still like Cheyenne Riley."

"I don't know. Well, maybe I have a different name in mind. I didn't get a chance to tell you, but I received a call a couple of hours ago when you ordered me to get something to eat in the cafeteria," Sam said, wondering if this was the time to lay some big news on his exhausted wife.

"What is it, honey?" Riley asked, already looking so much better.

How can his wife look this good, this beautiful after only a few minutes? She was amazing! Our little girl will be just like her mother, Sam thought, lost in amazement.

"Sam! What is the news? I can tell it is something important. What is it?" Riley demanded.

"Oh sorry," he said. "I guess I'm tired also. I got a call from Chip Scott, from the Selection Committee. We've been selected! For the *Calypso* crew."

This should have been the biggest news of their lives, but looking at his wife and newborn daughter, Sam knew it was not. She was! His new child!

"OK, Sam, what name were you thinking of?"

“I thought . . .” He paused and looked at his new daughter. “I thought we would name her Terra Nova. Since we are going to a new world, she should be named Terra Nova Winstead.”

“Hmm,” Riley thought for a moment, then said, “OK, I agree!”

Riley looked at her baby and asked, “What do you think of your name, Ms. Terra Nova Winstead?”

“She’s a special little lady, don’t you think, honey?” Sam asked.

Sam looked at Terra and then at his wife who seemed suddenly lost in very deep thought.

She turned to Sam and said seriously, “I think she might be a little more than that.”

CHAPTER 10

The Leading Edge of Destiny

"TERRA, HAVE YOU finished problem one yet?" Elias asked.

"Yes, there are three ways that light can travel from a source to somewhere else—directly through empty space; through a medium, like air or water; or by reflection from a mirror, for example," Terra answered.

Terra was younger than Elias, who was seventeen years old. Both were being tutored in person by Miles Kelly, a member of the educational staff and a physicist.

Terra excelled at math and physics. It was obvious to Miles that Terra had grasped all the concepts of optical physics. She had been asking for increasingly challenging practical problems. Miles felt that he could not challenge her enough with conceptual problems and was constantly trying to dream up practical problems that required mathematical solutions. At age ten, she was already at the eleventh- and twelfth-grade levels for math and physics, which was why she was studying with Elias, a junior in high school. The next logical step was to start her in college-level courses, which meant she would be in classes by herself as there were no college-aged kids on board who were awake at that time. Their virtual teaching assistant, Margaret, who appeared as a holographic image, agreed with Miles's assessment. Margaret was there to help Miles with the teaching strategies.

Miles, being an expert in the subject matter, was comfortable with the technical content, but teaching was something with which he and most subject matter experts needed assistance with. With all this help, she would be ready for college courses at the start of the next semester.

Later that day, as Terra sat in her quarters with Margaret, she began her final high school class, "The Grand Unifying Theory of Physics." Terra found this satisfying. To see how all the disparate theories of nature's forces were brought together in seven equations gave her confidence in the last year and a half of her studies. It all made sense.

Still, she did have ten homework problems to finish. She got through the first eight without getting stuck. The last two seemed to make no sense. Terra did not yet want to ask Margaret for help. She had set up each problem by stating the basic principle or theorem but could not seem to get the correct answer. As she struggled, which was not something she was accustomed to, she became frustrated, throwing down her holographic pencil in anger. She sat down on the couch in the little office area of their living quarters. Her mind started drifting. She turned on the outside star viewer. This always brought comfort to her.

As the stars came into view on the wall screen, she sensed there was something off. The star field was completely different. It was like she was viewing the stars in a different part of the galaxy or perhaps another galaxy. She sat up straight as she puzzled over the stars.

"Where is this?" Terra said aloud.

"It is the star field as it is seen from Mu Arae, your new destination and my home star," came an answer . . . but from whom?

It wasn't Margaret, who had suddenly disappeared. Who had answered? Now, Terra was shocked and frightened. She stood and started to back out of the small office, only to find that the door was shut and would not open despite her growing desperation.

"Don't be afraid, Terra," the same voice said.

"Who are you?" Terra asked.

"I am Eve of Eden, The Mother."

"Mother?" Terra asked, confused.

"Yes, I am The Mother of your home world, the Earth."

Terra felt dizzy, unsteady on her feet.

"I don't feel good, Mother," she croaked; then she fell onto the couch. Asleep or unconscious, she would not be able to say when asked about it later.

Terra dreamed. A dream she would have never dreamed. There was a landscape in front of her. It was a barren desert, with hills and mesas. The sun shone weaker, it seemed.

"The desert is supposed to be warm," she said in her dream voice, which came out much slower than it should have and with no breath behind it, making it just a whisper.

On the horizon, at the top of the nearest mesa, she saw a line of . . . animals? Or were they people? She walked toward the mesa; then she was *flying* toward it. She landed on the side of the line of people.

Yes, they are people, not apes or some other animal, she thought.

"Who are these people?" she tried to shout, but in her dream voice, she knew they did not hear her.

"They are your first fathers and mothers," The Mother replied.

"What does that mean?" she asked weakly.

"They are the first men and women of your kind on the Earth."

Terra looked at the haggard and dirty faces of the men and women. They did, in fact, look human, and there seemed to be an intelligence or "wisdom" in their stares.

They walked in silence until the man at the front of the line held up his hand and they all stopped.

They were wearing animal skins, but some of them had other clothing underneath, which did not match the animal skins in any way—light-blue jeans, with shirts of the same color. It would have been no leap of imagination to conclude that they were "cavemen" except for these clothes. They did not seem to fit at all with the outer garments.

"Why are you showing me this?" Terra shouted angrily at the sky, assuming The Mother was in the sky. "What does it mean?"

"Because, my child, you are to be The Mother of your new world. Ask your father about the message," The Mother said.

As she said this, Terra felt someone shaking her shoulder. It was her father, kneeling next to her on the couch.

"Terra, wake up. Are you OK?" he asked.

Terra fought to push away a deep fog of sleep as she whispered weakly, "I'm OK. I just had a weird dream."

She sat up on the couch, turned to her father, and asked, "Daddy, what is Mu Arae? Where is it?"

Sam looked at her but could not bring himself to give any answer.

"Daddy, tell me about the message."

Sam sat down on the couch next to her, carefully measuring his next words.

CHAPTER 11

Secrets Shared

"SO," SAM SAID, "how do you know about these things?"

"A woman, The Mother, told me," she said.

"Who is The Mother?"

"Tell me about Mu Arae and the message first," she said defiantly.

Sam was taken aback by the assertiveness of his ten-year-old daughter.

"A few weeks ago, when you were with me, we received a conventional message. I sent you back to our quarters, so you never heard what it was," he said. "The transmission, the part I heard, was music. Very unusual music. It seemed like it had control over me. It was difficult to turn it off. I didn't want to turn it off. I checked the origin point, and it came from the Cervantes star system, which is also called Mu Arae."

"What did the message say?" Terra asked.

As Sam explained the message to her, she sat, transfixed by his every word. When he was done, she asked her father if she could listen to the message. Sam agreed and asked her to come with him on his next shift.

It was now Terra's turn to talk about her dream. As she spoke, Sam sat in stunned silence.

"Terra," he began, "was this the first time that you had this dream?"

"Yes."

They discussed the meaning of the dream and who the people were that she had seen. Neither of them was entirely certain of the meaning, but both agreed that the people she had seen were their distant ancestors and that somehow, they were connected to The Mother's people.

Later, at the start of his shift, Sam and Terra walked back to ComCent. As the lead communications engineer and having been granted access to the classification level of the message, Sam could listen to it and, more importantly, he knew how to get to it without leaving a trace that it had been accessed. After retrieving the message, Sam played it for her to hear.

As the music played, Sam noticed that Terra began to smile and hum. Soon she began moving, swaying back and forth, and then she began dancing a complex motion that Sam had never seen. Finally, she began to sing in a strange language.

Sam sat transfixed by the sight of his daughter dancing and singing, with the music of the message playing softly now. Overpowering joy filled him, and soon he began humming then dancing in step with Terra. As they sang and danced, a beautiful feeling filled their souls.

Soon, an image formed in the center of the room. It was a woman, a beautiful, mature woman, wearing nothing but a light blue-and-white robe that opened as she walked toward them, revealing her legs up to her thighs.

Sam and Terra stopped singing and dancing as she began to speak in a language they had never heard, yet somehow they understood perfectly. Not only did they understand her words, but they also *felt the emotions she felt.* They were filled with love for her, yet also sadness for she spoke of an existential threat to a great civilization, her civilization. She told them there had been war. They were a peaceful people who had long ago given up all violence, yet in doing so, they had lost the ability to defend themselves. Hate, anger, and all other negative emotions had been expunged from their thinking, from their DNA, *and now she was pleading for their help.*

More images filled the room. A large group of The Mother's people could be seen living in a cavernous room carved out of rock. Men, women, and children were included in the group of people. Sam and Terra could feel the grief and sadness of these people. They had all suffered and were struggling with their heavy losses. They all looked at Sam and Terra and began to sing a new song, one that filled Sam and Terra with sadness that was so deep and devastating that both fell to their knees and began sobbing uncontrollably.

Eventually, Terra spoke through her tears to the woman, whom she now recognized as The Mother from her dreams. "Mother, tell us what you desire of us."

"My child, we need your help to save us as well as your people. We need you to fight with us. My people do not want to fight, but they will. They are not warriors. Your people will fight more easily and more expertly. We communicated our feelings to our enemies, but they refused to feel. They could not be affected as you have been by our message to you. We need you and your people to help us defeat them. We know what we are asking of you. Some of you will die violent deaths, but please know that the death of your corporeal bodies is not your end. Our *Souleng* or *soul energy* cannot be ended. We are eternal. We will come back as many times as necessary to empty our souls of hate, anger, and all emotions that lead us to harm ourselves or others," she said. "I will speak to you again when it is necessary. The time will come to share with all your crew what it is that I have told you. That time is not now. Tell no one else what I have said. Your lives and ours depend on you, Terra. You will lead them in this fight.

"Terra, there are a few things I need to tell you. First, do not sleep in your hibernation tubes. Do not worry about aging and dying on your voyage. You will not age beyond your thirty-fifth birthday. Secondly, I will teach you of our ways. Lastly, you can summon me by singing the song you just sang. Do this regularly, at least once a month as you measure your time. I will speak to you again the next time you summon me. My discussions with you will be very important to the success of this mission. Always seek me out and let us commune regularly." And with that, the image in the room vanished.

Sam closed the message and hid his digital fingerprints. He and Terra sat in silence for several minutes. Eventually, Terra left her father and went back to her school class, for which she was now late.

CHAPTER 12

An Attempt at Interstellar Insertion

Boulder, Colorado
Three Years Earlier

PETE WILM HAD spent his Sunday afternoon and evening drinking himself into a stupor. His wife of twenty years, Carissa, had walked out the door that morning following their latest argument. This time, it looked like she wouldn't be coming back. As with his previous marital setbacks, alcohol was his choice of anesthetic.

When he clocked into work the next morning at the Boulder Universal Actuators Electric Motor Assembly Building, his head was pounding, and he was barely awake. Today, his assignment on the motor assembly finishing station involved, among other things, installation of the RFI motor seal for nozzle actuator motor unit number one for the upcoming *Calypso* mission. As he stared at the parts list, his vision blurred due to the pounding in his head. He squinted but could not clearly make out the part numbers and part descriptions required, so he winged it, falsifying the fabrication completion checklist.

* * *

News of the course and destination change was shared with the crew by Captain Simmons in a crew-wide broadcast the next day.

After weeks of careful iterative work, the optimum course and acceleration profile was determined by Mission Control. Course changes were important for an interstellar voyage, and this was mankind's first manned mission beyond the solar system. They needed to get it right.

A double-check of the flight plan was always done, at least when feasible. *Calypso's* stellar flight engineers had checked the flight plan prepared by the Earth controllers to ensure that the course change was calculated properly.

The course change prepared by Earth was a dramatic change from their original destination and required significant nozzle vectoring. This was the first time the giant main nozzles would undergo a powered move by the actuator arms attached to them. At the time of their departure from the asteroid belt, the tugs had positioned the *Calypso* in the correct orientation, and no nozzle vectoring had been required. Major vectoring was expected to be a rare necessity.

Traveling to a very distant point in interstellar space was not as simple as pointing the ship in the direction of travel. Objects in space moved and accelerated relative to one another, and each object had a different motion and acceleration. These moving celestial bodies also exerted a gravitational attraction on objects near them. Since the *Calypso* would be constantly accelerating, the acceleration or deceleration profile and their route of travel to their new destination was an important input to the flight plan.

The flight engineers had installed the new flight plan into the ship's guidance computer over the last week. To not eat up the nozzle vectoring system's (NVS) lifetime limit of one hundred nozzle movement cycles, it was decided to do "dry nozzle movements," meaning the ion drive would be shut off. The nozzle movements would be "time zero" and "step one" for the new flight plan. It would need to be moved within a ten-minute window to keep the flight plan accurate. This window included time for two attempts. Earth controllers had informed the *Calypso* that failure to make the course change by timestamp 20:00 Earth GMT would cause the trip duration to extend by more than seventy-five years. This meant that moving them within the time limit was crucial. To that end, the *Calypso*'s flight control and propulsion engineers had prepared and practiced the nozzle repositioning movements, including preparing for problems.

Bob Wright and Seth Williams, propulsion technicians, were sent topside to the surface of the asteroid, to be ready if the nozzles failed to respond to the movement commands. A manual method for moving them was an option; it involved cranking a wheel, but the gearing on the crank wheel meant it would take about four hundred rotations of the wheel, which would take at least seven or eight minutes of hard manual labor.

Bob, the senior of the two technicians, hated being topside. The bulky spacesuit made work exceedingly difficult. The suit had shielding to protect against harmful solar radiation, which meant that movements that were simple without the suit became problematic with it. Seth, on the other hand, loved being outside of the living quarters.

"Seth, hand me my cheater bar," Bob said, sweat appearing on his red undergarment. The cheater bar would be necessary to break the wheel free if they needed to perform the movement manually. This could be a dangerous operation as there would still be pressure in the actuator hydraulic system, and that meant the wheel could spin once the lock was removed and the wheel was broken free from any friction holding it stationary. No one had told Bob how fast the hydraulic pressure would spin the wheel. The design of the system prevented the system from free spinning. Allegedly, the force of one man holding the bar to prevent spinning was all that was required. This operation, manually moving the nozzles, had not been attempted in system-wide pathfinder tests that occurred prior to the departure of the ship. It was deemed that performing a manual movement was not likely. This fact was not lost on Bob, a natural worrywart.

He was fairly sure that Seth had not thought of this; it was, after all, not part of his easygoing nature. He thought about this for a moment, concluding that he could have avoided a great deal of stress in his life if he could just not think about all the worrying possibilities that came with his job. Still, being a worrywart meant that he had never had a workplace accident. He was always well prepared, at least mentally, before he attempted something dangerous. And this operation, if it became necessary to perform, was full of danger.

In the control room, Engineers Andrea Lutz and Stephanie Pokorny were monitoring the readiness of the two topside technicians. At consoles to the left of them, a gaggle of engineers were working, monitoring the countdown to the ten-minute window, which was now only six minutes away. Execution of the new flight plan would begin then, with the powered nozzle movements being the first command. The backup plan, manual moving of the nozzles, was not considered likely, hence much more attention was being paid to the automatic nozzle movement. Everyone could see the flight plan steps live as they became displayed one at a time on the giant screen at the front of the room.

"T-minus thirty seconds, proceeding to automatic execution," the computer voice announced. The countdown continued to zero. At that point, all attention turned to the large screen to see what they all expected for step one: *Nozzle repositionings completed.*

Instead, the screen said, *Nozzle repositionings failed.*

Instantly, one of the engineers typed "Program halt," manually entered the new nozzle positions, and pressed Execute on the command console. This time, the command console read out the following: *Execution failure*. Groans could be heard along with some cursing from the engineers.

Captain Simmons turned and spoke to Andrea and Stephanie, who had remained bent over their console, and said, "Execute manual nozzle movements."

Andrea spoke into her headset, "Bob, execute manual nozzle movements."

Topside, he responded, "Copy, executing manual nozzle movements." He reached to depress the UNLOCK safety switch on the wheel, all the while never releasing his firm grip on the cheater bar.

What happened next happened quickly in a blur. Bob lost his grip on the cheater bar as the wheel began spinning. As the wheel spun, the cheater bar came loose and flew straight at his face shield. It struck his visor, which cracked, and lodged itself deep in his skull. Seth saw a stream of gas leaking out of Bob's helmet. He knew his friend had been instantly killed by explosive decompression.

Bob's face wrinkled. His blood boiled in the vacuum and shot from his face, escaping into space. The odd thing was he remained standing, with the bar impaled horizontally in his head. Seth did not understand why his friend remained standing.

Stephanie screamed as their video monitors showed what had happened. Andrea, remaining calm, ordered Seth to jam his cheater bar into the wheel spokes to slow the wheel down.

Seth crouched down; his plan was to shove the bar upward through the spokes in hopes that the bottom of the bar would jam itself on the rocky surface of the asteroid and force the wheel to stop. This worked, but not before the bar was significantly bent by the force of the spinning wheel spokes. At that point, he felt confident that reinserting the lock would hold against what was clearly an unexpected amount of residual hydraulic pressure in the NVS. He inserted the lock, and the wheel slowed to a stop.

He fell backward against a boulder and sat motionless for a full minute before he heard Andrea say, "We are sending out a recovery team, Seth. Remain in place until they arrive, then come back in."

CHAPTER 13

Course Change Achieved

CAPTAIN SIMMONS ANNOUNCED the news about the loss of their crew member the next morning. Everyone observed a moment of silence and then was told there would be a memorial service for Bob the next morning at 10:00 local time.

The memorial service was overseen by the Unitarian Universalist chaplain, Spencer Oaks. The service was simple and brief, in line with the belief systems that evolved with the collapse of organized religion in the twenty-first century. In the US, most organized religious communities slowly died as their tax-exempt status was revoked in the aftermath of the new US Constitution of 2047. What remained of the religious landscape was a simple spiritual movement based on tenets borrowed from Christianity and Buddhism. Many adherents of Islam and Judaism died with the nuclear exchange of the 2030s, which effectively wiped out Israel and the Muslim nations of the Middle East and Southwest Asia. The extinction of organized religion led to the rise of spirituality as society needed some sort of belief system to act as a moral compass. As a result, Bob's service was more a "celebration of his life" than the typical twentieth-century Christian funeral services.

Spencer began, "Our friend and shipmate, Bob, is no longer physically with us, but his spirit lives on in the realm of souls. In our memories, he lives on also, especially with his wife Kathy and his son Robert Jr.

"Bob was a devoted family man and a loving husband. He was a selfless soul and could always be counted on to help those who needed it. He was one of the first among our crew chosen for this mission. We all knew our participation in this great endeavor for humankind could be dangerous. Bob, and his friend, Seth, volunteered to perform this dangerous work topside. We fervently believe his soul has returned to the world of souls where he awaits rebirth in a new opportunity. We now commend his mortal remains to space, where they will be recycled by the cosmos. Bob, we bid you farewell."

As the weeks passed, the incident investigation revealed the root cause of the disaster to be a failure of an electronic circuit caused by installation of the incorrect RFI seal in the nozzle actuator motors.

With the replacement of the nozzle actuator electronic circuits and RFI seals, the nozzles could now be repositioned electronically. The manual nozzle movement failures, which had killed Bob, remained unresolved, but it was decided to abandon the procedure altogether as it was viewed as unnecessary given the obvious root cause and fixes that were found and implemented.

Seth, Andrea, Stephanie, and the leadership team were all deposed as part of the incident review. The preliminary cause for the failure of the NVS was burned-out electronics, topside, on the actuators. This likely happened when the *Calypso* had encountered streams of highly energetic particles from a coronal mass ejection from the sun. The course of travel to that point had put Calypso within 0.5 astronomical units (one-half of the distance of the Earth's orbit to the sun). This meant the ship had been exposed to much higher fluxes of energetic particles than it had in the asteroid belt. The protective enclosures on the electronics had been improperly installed, leaving them vulnerable to the solar environment. The electronics were replaced, and new and beefed-up protective containers were fabricated to protect them from any further similar incidents.

This left the Earth and the crew with the sobering reality that they would need to wait for a new flight plan and then try the whole thing again. While they waited, the flight engineers reignited the ion drive and continued to their original destination. Another attempt would come when they were on the other side of the sun and farther away from it, just to be safe.

The updated flight plan arrived in a noticeably brief time, four days later, and was checked by the *Calypso* engineers. This time, the nozzle vectoring would occur at approximately 1.5 AUs, which was deemed to be safe even with no protection on the electronics. Given that the protective enclosures had been installed properly, there was essentially no concern this time. At the appointed time, the flight plan was initiated with the nozzle movements and the course correction was made. Their new destination in the Cervantes star system awaited, 1,250 years in the future.

Once the interstellar insertion was completed, the awake crew would be cut significantly. The captain, senior engineering and scientific staff, and some other support crew would remain awake for another five years. Sam, his wife Riley, and Terra would remain awake for another ten years as those with children needed to be awake to raise their offspring. This meant that Sam and Riley would be older than most of the crew when they arrived at their destination. The math was a bit complex, but no one would be over sixty years old when they arrived. This included those with children who would be about ten years older than those without. Given that the human lifespan was about double what it had been in the early twenty-first century, sixty years old now was the equivalent of thirty years old then.

For now, Sam did not have to invent a reason for his daughter to stay awake the entire duration as The Mother had told them. The captain would certainly want to know the reason, and Sam was not sure what to tell her or his wife, who also knew nothing of the message and other details that went with it. Most of the crew believed their course change was simply that a more likely habitable world was found than their original destination. At this time, Sam did not need to do or say anything. He also did not know what else The Mother might share with Terra and him. He was in a state of not knowing the details or even the big picture in any real sense.

CHAPTER 14

What Was That?

INTERSTELLAR TRAVEL INVOLVES travel through space that is unexplored. Many objects in this expanse had not been discovered and charted, such as comets, asteroids, and even rogue planets that did not orbit any star. In the early twenty-first century, an unusual chunk of rock, shaped like a cigar, entered our solar system and quickly transited it. Its origin remained unknown.

For reasons such as this, the *Calypso* was in the business of identifying and cataloging any objects that they came across. Six months had passed, and the *Calypso* was accelerating nicely. However, a collision with any object, depending on its size and relative velocity, could be disastrous. To prevent this, the *Calypso* had radar directed in all directions to detect any nearby objects, and it had various telescopic devices to detect objects farther away from the ship. Astronomers, who were members of the scientific staff, monitored this automated system. Every detected object had to be evaluated to determine if it posed a threat to the ship.

Dr. Carl Carroll, Stuart's brother, was the senior astronomer and had three junior astronomers working for him in this effort. Dr. Manu Patel was one of these individuals. Manu was working the Saturday graveyard shift, looking through the automated cataloging observed in the previous twenty-four hours. He was also noting and reviewing any objects found during his shift, performing a real-time analysis. He had set up the program to alert him of any nearby objects while he reviewed the listing of objects found in the previous twenty-four hours.

At 01:15 hours' ship time, the console beeped. To Manu's surprise, the red light on the wall began flashing and an automated voice announced, "Collision threat, urgent response required." This was as serious as it could get, and a well-scripted protocol automatically alerted the captain and ship security/defense officer (SSDO), Gunner Lawson, a former space force pilot, both of whom were in bed at that time of night. Within two minutes, the two were in the observatory control room with Manu. The captain had donned her uniform; Gunner came running into the room in his boxer shorts and T-shirt.

Captain Simmons immediately asked, "Time to impact (TTI)?"

Manu responded, "Six minutes."

Gunner had taken the seat at the ship defense system (SDS) console next to Manu. Gunner responded that the TTI, size, and velocity mandated that the rail gun be used to take out the object. Without input from Gunner or anyone else, the SDS would determine the appropriate countermeasure. In some cases, this would be small bursts from the ship's pitch-and-yaw-maneuvering rockets to change course ever so slightly to avoid a distant object. If the object were much closer, and an imminent threat to the ship, the SDS could take multiple actions, such as adjusting course slightly or employing the rail guns to either destroy or deflect the object. The object in question was far too close to make a course change in time with the pitch-and-yaw engines.

The SDS began firing the rail guns at the object, now only three minutes away. The object kept coming, completely undeterred. This baffled the captain and Gunner.

Captain Simmons spoke into her headset to Bart Jensen, the engineer on duty in the guidance and propulsion room. "Emergency pulse from the ion drive, STAT!"

The three were thrown off their feet, landing in a heap on the floor as the ship's acceleration suddenly increased. Picking herself up, the captain shouted at Gunner, "Status? Time and distance!"

Gunner responded, "Three thousand meters, it's passing our stern now!"

The captain exhaled deeply and audibly.

Gunner continued, "We're in the clear, captain."

The captain's wrist communicator chirped with incoming messages from around the ship wanting to know what was happening. Captain Simmons tapped her ship-wide comm button on her wristband and spoke evenly, "This is the captain. The situation is stable. We accelerated to miss a nearby object. All sections, status report, damage and injuries." The captain seemed relieved, but Gunner and Manu continued to look at the screens showing the object's course and speed.

Gunner turned to the captain with a perplexed look and said, "Captain, the object accelerated as it came within ten thousand meters . . . how could it do that?"

Captain Simmons was caught flat-footed and turned to Manu, asking, "Manu, why did it suddenly accelerate?"

Manu fumbled for words. "It must be an instrument error, captain."

The captain and Gunner were unconvinced.

"Run a system-wide check," said the captain. "Get back to me with the results."

Manu turned back to his console and initiated a system status check. He muttered under his breath to no one in particular, "It better be an instrument error. The alternative is not believable."

CHAPTER 15

An Abundance of Caution

"IT LOOKS LIKE it accelerated toward us," Manu said to Carl, his boss, the next morning. "I ran the diagnostic, twice, on the radar and looked at the plot of its velocity, which increased as it closed in on us. It accelerated."

Carl held the plot of velocity and raised it up to put more light on it, studying it as he did. "I have no explanation except that it had its own propulsion and guidance, which means it wasn't a space rock, it was a weapon of some kind," Carl speculated. "Let's discuss this with the captain and Gunner. We need to jointly assess what this threat is and what we need to do to mitigate it. For now, let's keep this quiet. No use getting others worried, at least not yet."

Later that day, the captain, Sam, Peter Chan, the Carroll brothers, Gunner, and Manu met to discuss the incident and the findings.

"Captain, the object from last night began accelerating when it came within ten thousand meters. When it got within about three thousand meters, it began jamming our radars. One of our projectiles scored a hit on it at one thousand meters, sending it spiraling wildly, causing it to miss us. The emergency pulse of the engines threw a massive cloud of ions behind us, obscuring its radar, saving us from giving it another go at us. In short, we escaped because of improvisation. We really don't want to be in a toe-to-toe fight with these things again. I keep coming back to two questions. First, how did anyone know our position close enough to give this weapon a shot at us? Secondly, who would even want to do this to us?" Carl paused and looked around the room as he finished.

Captain Simmons, Stuart, Peter, and Sam exchanged glances at one another, the unspoken question written on their faces, but unknown to the others.

"What? What is it that you are not saying? Are you not sharing something with us?" Gunner asked the captain.

"It's probably not relevant to this incident. I will share some things with you when I can. If it were up to me, I would share it," the captain responded.

Sam started to say something but stopped himself, drawing in a deep breath as he did so.

"Sam, do you have something to say?" the captain asked.

"No, not really. Well, maybe . . . no, I don't," Sam stammered, not helping his believability.

Captain Simmons stared at Sam for a moment, trying to gauge his facial expression.

"OK, Sam, if you change your mind, you can always come and talk to me."

“I sure hope none of these secrets you all have are relevant to what I see as a clear and present threat to our ship, our mission, and our lives,” Gunner said exasperated, looking at Carl and Manu who nodded their heads in agreement, with puzzled looks on their faces.

Finally, Captain Simmons stood and declared they needed to end the meeting, but before dismissing everyone, she said, “No one outside this room is to be told about any of this information. Carl, I want you and your astronomers to fine-tune your searches for objects like the one we encountered. Gunner, I want you to oversee what the astronomers do and report to me any relevant information. Sam, I want you to cease providing Mission Control with positional updates of the *Calypso.* I know that our EPC transmissions should be secure, but we don’t know if we might have an insider threat in Mission Control or perhaps even on board our ship. Does everyone understand their duties?”

“Understood,” everyone repeated in unison.

As she stood and walked to the door, the others stood and said in unison, “Aye, aye, captain.”

CHAPTER 16

Family Matters

TERRA AND HER friend Cassia were doing their history homework together at Cassia's quarters. Cassia's parents, John and Binna, whose ancestors were Korean refugees from the Wars of Aggression in the twenty-first century, were both working their shifts, leaving the two of them alone. At ten years old, both girls had a shared interest in the same boy, Hatcher. They did not know it; however, Hatcher shared an interest in them, but neither side could bring themselves to admit to any such interest at ten years of age. That would have to wait for a few more years, then such feelings would be impossible to miss and control. For now, the shared infatuation was a secret.

Cassia shared with Terra the disagreement her mom and dad had over her course of study and future occupation. Her father was a zoologist, and her mother was a botanist. Her father felt that Cassia should be trained as a botanist or zoologist as well. Her mother disagreed and felt that she should be allowed to pursue her own choice of vocation. Cassia shadowed her parents in their duties, but she didn't want to do what they did. Cassia wanted to be an electronics technician.

Technicians, and especially electronics technicians, were trained to go topside to do maintenance on the equipment. They also were likely to be on the shuttle crew when they arrived at their planet. Cassia wanted a more adventurous occupation; she could not imagine that performing botanical or zoological duties would ever give her the thrills she desired.

"Tina's parents go topside almost every week," Cassia told Terra. "They get to pilot the shuttle on the monthly exterior inspection."

Most technicians were also trained to pilot the shuttles that were used to do inspections of the asteroid's exterior, the propulsion nozzles, communication antennas, and other equipment topside.

Cassia sighed and said, "That is so much the bomb!"

Terra smiled at her friend's passion. Cassia was very extroverted, which Terra admired. Terra, on the other hand, was introverted and always obeyed her parents. She wished she had the same courage as Cassia. She consoled herself with the thought that each person had different abilities. Cassia was a risk taker, oblivious to the dangers inherent in many of the things she wanted to do. Terra always calculated risks and would often not do something if it were too risky. Terra decided that she and Cassia could not be more different, but somehow that worked and seemed to make their friendship grow more.

At that same time, in quarters that were as far away as possible on the ship, Cassia's mother, Binna, cried out in pleasure. Her lover, Alex, a woman almost ten years older, sat up, leaving Binna on her back.

Both women were nude and sweaty, having worked hard in their pleasure-filled activity. Alex was tall, almost 183 centimeters with toned muscles. She was a space force veteran like Gunner. She likewise worked with Gunner in ship defense and security. "I'm going to shower. Want to join me?" Alex asked.

"Sure." Binna smiled, got off the bed, and headed to the shower with her lover. Binna was bisexual, and this did not seem to bother her husband, John, Cassia's father. Binna allowed him his sexual trysts since she had her own lover.

Binna's work required her to tend to the food crops; this meant she worked odd and unpredictable hours sometimes, with downtime in between certain tasks. When she had finished adjusting the water drippers, she had to wait for several hours before she was needed again. This had given her the chance to call Alex to see if she had time for "extracurricular activities." Alex always had time for sex. Binna was almost certain that Alex had the biggest libido she had ever run across, and Binna knew a thing or two about libido and sex in general. She had paid for many of her wants as she studied at the University of Zurich through her work as a "pleasure provider." Binna was a beautiful woman. She was thirty-five years old, 170.2 centimeters tall, and weighed fifty kilograms.

Views on sex had evolved by the twenty-fourth century. The government gave all adults unlimited sex passes to be used with pleasure workers. This progressive view of sex brought about a lowering of sexual tension in a society where marriage rates were less than 20 percent and reproduction was strictly regulated to one child. Every adult, old or young, beautiful or not, had the ability to satisfy their sexual desires through pleasure workers, and the government paid for it. There was no stigma attached to sex or to those whose job it was to provide sex. This liberalized view on sex had been adopted as a compromise to the mandatory pregnancy prevention provided by the chip implanted in all babies at birth. As a result, the world population had declined to four billion people. Nuclear wars and famines from global droughts had a hand in population reduction as well.

After the two women finished their showers, Binna dressed quickly; she did not want to be late for the afternoon inspection of food crops. While the agricultural robots had taken over most of the menial labor, inspections by human beings often detected crop issues not detectable by robots. Food production was of critical importance to the survival of the ship's crew and mission.

Later that evening, Cassia and Binna sat at the dinner table. Cassia's father, John, was working. Cassia picked at her vegetables, her head held up by one hand under her chin.

"Mommy, I really want to be an electronics technician. I want to shadow Maria, please, Mommy!"

"We've been through this, sweetie. Your father isn't convinced yet. He thinks you should give botany some more time," Binna said.

"I can go to the family judge and tell her this is what I want!" Cassia firmly said. "She can give me a skills-and-desires test. I know it will show that I would be a good tech,"

"Cassia, I am on your side," Binna said. "Let's let your father reach the same opinion as us, honey. I know he will."

Cassia sighed and nodded her head.

CHAPTER 17

A Talk with Mother

SAM HAD BEEN troubled ever since he and Terra had been overcome by the message of The Mother. He knew his daughter was supposed to speak to The Mother in regular intervals and that it would come by relistening to the original message. And therein was the problem. To summon The Mother, they had to play the beginning of the message and listen to the "music"; at which point would they begin "dancing" again and "singing"? He was a bit fuzzy, to say the least, about how to do it.

A month had passed since their first encounter. Terra and Sam were unsure if they remembered hearing "monthly" as the interval time. Additionally, accessing the message was dicey. Eventually, someone would find out that the message was being retrieved and listened to. There had to be a better way to summon The Mother. To that end, Sam intended to record the message, thus freeing them from listening to the message in ComCent.

He planned to do it when he picked up a graveyard shift, which would be starting this Monday evening. He told Terra his plan, and she agreed with the idea.

When Monday at 11:00 p.m. ship time came, Sam was sitting at the console. The ship was quiet. He assumed the captain and others, even the ever-present Gunner, were asleep.

There isn't any nightlife on the ship. Most people are the early-to-bed and early-to-rise types, he thought.

Sam went to the secure file location and entered the password that only he and the captain knew. He readied his recording device and opened the file. The music began playing. Sam found himself overcome by it, as before.

It took a moment for Sam to feel the tap on his shoulder. He turned and saw a large African American man with perfectly white teeth grinning at him. It was Gunner. He was shouting at Sam above the impossibly loud music. Sam turned down the volume and tried to grab his recording device so Gunner would not see it, but Gunner saw it and grabbed it before he could.

"What is this, dude?" Gunner demanded.

Sam paused the message, stopping the music.

"Um, . . . ah, it is a New Day music group on Earth. Terra wanted to have a copy for her playlist," Sam stammered, quite unconvincingly he thought.

"Right," Gunner said. "Well, since you are the boss here, hoss, I think I should let you know El Capitan dear has asked me to keep an eye on things down here. Not sure why, but she stressed that only those with access should ever be in here."

He continued. "Since you are clearly authorized to be here, *no problemo*. But you know, I am a suspicious person, I get paid to be, so I'm watching everything, just in case. I mean, the captain must have had a good reason to tell me to watch things down here. I know you are not supposed to be listening to anything non mission related, so I'm going to let things slide this one time, but I do need to hear what's in the message, dude."

"Sure, OK, ah, . . . let me restart it," Sam said, his voice quivering.

Sam felt certain he was in trouble. His hand shook as he moved the cursor back to the start. The music began playing. Sam looked guiltily down at the floor. Within seconds, he found himself transfixed by the music and noticed Gunner was also. The music continued playing with both staring blankly ahead at nothing, their bodies swaying slightly. At this instant, The Mother appeared in the room, looking the same as Sam remembered. The music continued playing, but the volume was now quite low, letting the music fill the background softly; its effect on both seemed to diminish greatly.

The Mother spoke. "Gunner Lawson, listen carefully. Your assistance is needed."

Gunner's knees failed him, and he hit the ground with an audible thud.

He stammered, "Whaa, who are you?"

The Mother seemed to have sensed his question beforehand as she answered immediately, "You will learn who I am shortly. For now, you are to give any required assistance to Terra, Sam's daughter, and Sam. They have important work to do on behalf of the Earth and *my people. My people* need help from this ship and the people on it. *Your world desperately needs help to avoid extinction.*"

Sam instantly noted that there was added information in her words this time. He also saw that Gunner, who was still on his knees with his head cocked to one side and his eyes intently staring at The Mother with his mouth open, said nothing. Sam smiled slightly.

The Mother continued. "Sam, you will call me by playing the music from your recording device. I will let you know we need to talk by making the rock in front of you change color from black to white. Sam, the rock is white. Go to your quarters. I want you, Terra, your wife Riley, and Gunner to be present for this new message. Sam, call your coworker Neil to come and perform your duties this evening. Neil will not ask you why. He will do it without questioning. Then listen to the message recording again. As you begin singing, I will come."

Sam had to help Gunner up and hold his hand all the way back to Sam's quarters. He sat Gunner down on the couch and told him to stay put; then Sam went to his bedroom to wake his wife.

"What time is it, Sam?" she mumbled.

"Uh, it is very early. I have something particularly important to show you. Please, come into the living room."

Sam continued to Terra's room even as his wife struggled to put her feet on the floor and slip on a bathrobe.

In the hallway, he nearly bumped into Terra.

"Terra, honey . . ." Sam stopped as Terra spoke.

Terra was standing, fully dressed, and cut him off. "I know, I'm coming. The Mother summoned us."

"Y-yes," Sam stammered, confused that she was up and somehow knew what was happening.

As they gathered in the living room, Sam told Gunner, who did not yet know, that some time ago, a message had come in that originated on a planet outside their solar system. He asked that they hold all questions until after they listened to the message.

As he pressed PLAY, the strange music began, only the quality of the sound far exceeded the capability of the cheap little device he had. Each of them became mesmerized by the sound of the music, and shortly they all began swaying, staring blankly ahead at nothing. Then they began humming to the music, and finally, they began dancing in unison.

The Mother appeared. They stopped humming and dancing and stared at her.

She smiled and addressed each one by name. "Terra, Sam, Riley, and Gunner, I asked all of you here because I have important things to tell you. Sam and Terra know me as 'The Mother.' My people and I reside on the first planet that orbits the star you call Cervantes or Mu Arae. My people are under siege by an enemy who is not pacified *by our thoughts and feelings.* We are in a position of stalemate and have been for hundreds of years. They have retreated, but we know that they will return in the next thousand years or so. Our nature is not vicious enough to defeat them, and because of this, we will not be able to perform our work for The Ones, those whom we all serve. We need your help in defeating an enemy who knows of the existence of your home world, the Earth. They must defeat us to defeat you. We have protected you and nurtured your existence for many millennia. Your captain, crew, and Earth do not know of this conflict and our need for your help, and we need to keep it that way until the time is right. Those who know of this message only know that we have a habitable world for you in our star system, and that is why your destination has been changed. We will eventually share valuable information with you to help you to help us in defeating this enemy.

"Terra will become The Mother of your new world, and she will lead the fight against this enemy. Please give her all the support she needs. You need to know that we created your species for a purpose that you do not yet understand. I am your 'Mother'. We have always loved and cared for your world. Our intentions are simply to continue to help your species reach its fullest potential. Be strong and do not waver. Good will prevail if you do so. For now, I bid you farewell. We will speak again as it becomes necessary."

When she finished, The Mother disappeared, the message stopped, and each of them was left to their own thoughts.

After a couple of minutes, Terra stood up and thanked each of them for coming to listen to the message. Her words and mannerisms seemed to have changed, making her appear much older and more mature than the ten-year-old girl who stood speaking to them. None of them realized it at the time, but they looked at Terra in a completely unique way now.

CHAPTER 18

The Poker Table

TWICE A WEEK, Captain Sandra Simmons, chief engineer and second officer Peter Chan, Gunner, and Alex played cards. Friday night was bridge, and Tuesday night was poker. The games were competitive, to say the least, as the people playing were extremely competitive.

Sandra was a graduate of the air and space academy in Colorado Springs. At thirty-one years of age, she was the youngest of the four captains of the ship. Three others were in hibernation. Ranks in the space force more closely resembled naval ranks from the twenty-first century. A captain was the equivalent of a colonel in the air force, making a space force captain the equal of a navy ship captain. She was quite young to hold such a rank, and she was recognized as the most competent officer among the captains on board, making her the lead captain.

Peter Chan, the chief engineer, was an MIT graduate. He liked to remind people of this whenever required, which was often to those who knew him. It made him seem insecure to some, but it was unnecessary; Peter was a terrific engineer. He had over forty patents on gadgets that touched every field of engineering. He had been a member of the senior engineering team that designed the *Calypso* and had personally solved many of the most troublesome issues in what was one of the most complex engineering projects in human history. Peter's proudest achievement was that the *Calypso* was designed to function for over two thousand years, something that had previously been impossible. Sometimes reliability statements such as this sounded like marketing speaking, but most experts believed this was actually conservative. Taking all this together, Peter was a tad bit competitive also.

Gunner and Alex, who were both members of ship security and defense, were both unimpressed with Simmons's and Chan's pedigrees. They were both certain that mission success would come down to their skills and bravery.

Hubris abounded at the card table, so much so that it seemed unlikely they could all fit their egos into Captain Simmons's stateroom where the card game took place.

The game tonight was Texas Hold 'Em, and it looked like it was coming down to Peter and Gunner; their luck tonight knew no end. Sandra and Alex were bonding over their obviously bad luck this evening.

Alex stopped everyone's concentration when she asked, "Gunner, I saw you and our lead comm engineer head into his quarters last night. Do you and he have something going on the side?"

She smiled as she said this, knowing that would blow Gunner's concentration.

Gunner, startled by the question, immediately answered, "Of course not! The dude wanted me to help him with . . . ah, I can't remember. Alex, goddamn it, I'm trying to concentrate!" He shot back with a little more seriousness than seemed appropriate for the playful banter.

Sandra and Peter exchanged a look of surprise. Then Peter said, "Alex, give the man a break. He just broke up with Shelly Swanson. That heartbreak would bring any red-blooded man to his knees, but it wouldn't come close to having him switch teams!"

This disarmed the tension, and Gunner found himself saved by the bawdy humor.

Gunner thought to himself, *I need to watch my six. Maybe El Capitan asked Alex to watch things also? Mental note, keep an eye on Alex.*

The game ended with Gunner winning despite his momentary loss of concentration. As they rose to leave, Sandra asked Gunner to hang around for a minute.

"You haven't seen anything unusual going on in ComCent, have you?" she asked.

"No, nothing," he answered.

"I'm thinking we should take the message file off the secure server and download it to a data bit and put it into the captain's safe. Mission Control on Earth has asked me several times about how I am safeguarding the message. I think it would be prudent to do this," she said. "I want you to be down in ComCent when Sam pulls the file and saves it on the data bit. Make sure it is scrubbed off the server then bring it to me."

"Copy that captain," Gunner responded.

After he left the captain's quarters, Gunner noticed that he was sweating through his uniform. *Secrets*, he thought. *Never liked them!*

CHAPTER 19

The Enemy Within

"SHIT!" ALEX SAID as she tried a password in the secure file directory.

She had a piece of Sam's DNA for the bio-identification check, but the password step eluded her. A third unsuccessful try would lock her out of the system.

Alex stood and looked around. The room was still empty, the life-support "failure" was keeping people out. She figured she had about three more minutes before environmental engineers declared the room safe. Alex had no choice; the password she thought was the correct one was not. Emissions that she had picked up from this keyboard when Sam had logged in thirty minutes ago were not decipherable by her device. She figured it had been worth a try. Alex knew the captain had ordered the file removed. She had learned that from talking to a clearly drunk Gunner as they played darts after their shift. Gunner could be lubricated into revealing things when he became impaired; she knew because it had worked on another occasion when she got him to talk about a strange message.

Loose lips sink ships, a refrain from the twentieth-century war with the Axis powers. Would that have been the Second World War? Alex wondered. *Had to be. My history teacher in high school had referred to it as the "Just War." When was any war just?*

"They could be back any minute. Not worth the risk," she muttered under her breath.

Alex closed the password window and scooted out of the room via the ventilation shaft tunnel, the same way she had entered. She had tricked the sensor on the automatic emergency door in the ventilation shaft to think it was closed in order to get in after the pressure-loss alarm sounded.

It was a good try, she thought, but a dead end.

CHAPTER 20

Push Forward to Live or Stop and Die

East Africa, Earth

10,911 Years Before the Common Era

EVE OF EDEN, The Mother, was exhausted and certain she could not continue onward. She was nearing her limit. Her distended belly seemed to pull her to the ground. Her people had left the ice behind, but the snow remained. Their home, in what would become known as the country of Turkey thousands of years later, was now far behind them. She would give birth to her first child in a new land. But it was not an empty land. There were other hominids here, but they were different in particularly important ways.

They did not possess the sophisticated tools her people had, and they were very suspicious of outsiders. Eve's people were peaceful and welcoming to strangers. They possessed the peaceful mind given to them by the Lords of the Skies. They pacified many a violent neighbor with their peaceful and loving thoughts carried by Eve's telepathic abilities. They would need that here, as never before, if they were to survive.

"Shall we make camp?" Adorn, her husband asked.

"Yes, my one love," she answered. "We are past the ice, now there is only snow. We can make the Great Cracked Valley in a few more days."

Eve and her people were moving south. The Sky Lords warned Eve of a great destruction that was coming. Their warning foretold a time to come when a comet would fall from the skies, melting and breaking up the great ice sheets covering much of the northern reaches of the westernmost continent. The influx of freshwater into the northern oceans would halt the northward flow of warm southern water, causing worldwide temperatures, which had been rising, to (temporarily) plummet. The large inflow of fresh water from the melted ice sheets into the oceans would cause worldwide sea levels to rise by four hundred feet within days. Millions would die from the flooding of coastal areas, and many more would die from the disruption of the food supply. This event, in modern times, came to be known as the Younger Dryas Climate Anomaly.

Eve was to preserve the knowledge of the Sky Lords by helping herself and her people survive the coming cataclysm; after all, they had vast knowledge given to them by the Sky Lords. Moving south would spare her people from the catastrophic floods that would overfill the great Mediterranean Basin and spill over the

Bosporus Strait into the Great Depression, north of their home in Eden.

Elotim and the other Lords had left in great haste, but they had not left Eve's people helpless or hopeless. Elotim told them where safety and food could be found, far to the south of their beloved Eden, in the Great Cracked Valley, the original home of all upright peoples.

Eve was the most educated of all her people. Mareye, beloved wife of Elotim, had borne Eve and had taught her all the ways of the great Sky Lords and their history. Eve had been told from her earliest recollections that she possessed the traits of the Sky Lords and would pass them on to the children of the new Earth.

The Lords of the Skies had left them with supplies and tents for their journey, but they had no pack animals to carry the supplies. Instead, each carried what they could. As they sat around their campfires, the travelers ate dried meat and nuts. Abelom, Adorn's oldest child from his deceased first wife, sat hunched and cold next to the fire.

"The women are exhausted. I hope we make the Great Valley before we lose any more people," Abelom said.

"The Mother says we are only days away," Adorn answered.

"Why did the Lords leave so suddenly?" Abelom asked.

"They said the balls of fire from the sky were a danger to their great sky chariots and that they will not be back for an exceedingly long time. I am so sad to see them go," Adorn said. "They said we are now ready to be on our own. Perhaps we are like their children and are now grown up."

Kaanon, Adorn's second eldest son of his first wife, said, "Perhaps they lost interest in us and were weary of the primitive nature of our existence. The Mother saw the inside of their sky chariot and described wondrous things inside it. What do we have to offer them anyway?"

"They always told us we were their purpose, their 'reason' for creating our world!" Adorn shot back.

"If we are so important to them, why did they leave us at this dire time?" Kaanon asked cynically.

"Kaanon, speak respectfully of the Lords!" Adorn said, raising his voice.

"They aren't gods."

"Perhaps so, but we owe them so much. They created our world for us."

"Yes, a world that is still cold and dark and will soon be pelted by rocks falling from the skies!" Kaanon retorted.

"Enough! The Lords are gone for now, but they provided for us. Eve is our leader now. She knows of their ways and will sustain us."

Eve sat in her tent, bending over a small white stone. "Elotim, please show me the way from here to the valley. We are very weary, and some are complaining of the hardships we have faced."

In the middle of the tent, Elotim appeared, his pointed brain cap atop his head. “My dearest Eve, the way is straight south from where you are now, three days’ journey. By tomorrow near evening, you will find a path to descend from the tabletop mountains. A flat plain will lead you into the mountain pass and then down again into the great valley,” he said with great love apparent in his voice.

“Thank you, Father,” Eve responded.

Elotim blinked out of the tent. Eve sighed and lay down, the weariness in her young bones overwhelming. Within minutes, she was asleep.

CHAPTER 21

Making Friends

THE MOUNTAIN PASS gently descended into a broad, flat valley floor with a large lake in the center. The vegetation was denser and lusher. They camped at the north end of the lake where the river exited. Eve's people were skilled fishers and were immediately standing in the shallow waters spearing fish. A group of her people were digging diversionary ditches to bring water to the flat area they had designated for crops.

A small group of darker-skinned people, the Toba, watched with great interest as Eve's people dug long straight ditches between long linear mounds in which the seeds were planted. The Toba and the other tribes had never cultivated crops. Instead, they moved around the large valley and gathered fruits, nuts, berries, and fish. The Sky Lords had told Eve to share the knowledge of planting, caring for, and harvesting plants for food. Eve hoped that sharing these ideas with the local tribes would foster a kinship with them. It seemed to be working. The Toba and another tribe, the Tree People, were now moving to the area where Eve's people had gathered. They were showing Eve's people how to build huts from the local trees. The floors of the huts were elevated off the ground by a meter to account for the flooding that came in the rainy season. The local tribes spoke a variation of the original tongue that Eve's people still spoke though it was not always clear what they were saying as the language had "drifted." The original tongue had been taught to the upright peoples many generations earlier by the Sky Lords. The local tribes knew of the Sky Lords though they had corrupted many of the stories with curious changes to make their tribe seem central to the story of the Sky Lords.

As time passed in their new home, Eve slowly evolved into the leader of all the tribes. Her wisdom and knowledge of climate, agriculture, and tools earned her a position of respect. This was to the benefit of all the peoples as they now had peace. In former times, they had known serious conflict. Eve was a leader that unified all of them. The tribes did not know of Eve's ability to communicate directly with the Sky Lords. Elotim had told Eve to hold this secret from all but a few of her own people. The local people now referred to Eve as *Faroh*, or "Great Leader."

Eve had Adorn set up a large hut where the children would be taught in various subjects by Eve's women, who were experts in many subjects, including the symbolic writing method of the Sky Lords. Most importantly, Eve had told her teachers to instruct the children about the "purposes and methods" of progress through their lives. Eve knew that the people needed to know of the great

Wheel of Lives and of the need to grow toward love and caring of all peoples. They would be taught of the Great Source, which stood above all, including the Sky Lords. Some aspects of this had been retained among the valley tribes though time and a lack of a written language had caused the important principles to be corrupted and changed into almost unknowable gibberish.

Eve's firstborn son, Seff, now fifteen years old, was always at Eve's side. All regarded him to be the heir apparent to Eve when she passed. Seff looked like a younger version of his father, tall and broad shouldered. He was wise like his mother and had the same gift as her, the ability to read the thoughts of others and communicate with them telepathically. It was exceptionally rare that a male would inherit this ability. This was the last piece of the human genome that the Sky Lords wanted Eve's people to have.

It was believed by all that Eve would live to a great age. Her genes from her Sky Lord parents provided far more durability in replication than possessed by any of the upright peoples. Over the years, many of Eve's children passed; but she, Seff, and several of her children lived on, having inherited their longevity and telepathic abilities from the Sky Lords.

Adorn had long since passed, and Eve took other husbands and had more children. As she reached her four hundredth birthday, Eve left in the night and was never seen again. With Eve gone, Seff became Faroh in her place. The cold and ice abated in the world, and the descendants of Eve began leaving the valley. The land around the great lake was subsiding, resulting in warmer conditions and more rainfall. The increased rainfall caused the great river to increase in size and volume, leading to greater fertility in the lands to the north. This created new lands to be settled and cultivated. They became the ancestors of the ancient Egyptians.

Eventually, Eve's descendants went north and east to the great rivers in the land of Mesopot, where they became the distant forebearers of the great civilizations between the Tigris and Euphrates Rivers. The Sky Lords still had not returned, and people gradually forgot about them. The gift of telepathy and thought reading became rare among humans, but it never completely disappeared. War and conflict became more common, leading to the loss of many of the great pieces of knowledge that were once had among the human race. However, some of that knowledge persisted for thousands of years and led to many of the great feats of engineering, astronomy, and architecture that could be seen in the ancient world.

CHAPTER 22

Knowing, Seeing, Feeling

TERRA WOKE IN her bed on the night when the alarm in the engineering section went off. She felt a pull toward a person in the engineering space. She sensed urgency and desperation. Terra focused on the feelings. As she did, she started to see through the eyes of this individual. She saw ComCent and hands furiously typing on a keyboard, with the alarm blaring both inside and outside her head simultaneously. She winced as the sound became painful. It seemed as if it was strangely multiplied by two. The view changed as this person stood and walked to a chair that was set underneath a ventilation shaft of some kind. They climbed into the shaft and replaced a grate over the opening. The confined space of the ventilation shaft came into Terra's view as the person crawled through the shaft and reemerged in a hallway outside the engineering sections. Terra sensed their urgency, fear of being caught, and some guilt. The woman walked through the halls and stopped next to a group of people, which included Terra's father, Sam, Gunner, and the captain.

"Alex, where have you been?" Gunner shouted above the blaring of the alarm. "I thought you were monitoring these sections."

"I went to the kitchen space for some soup," Alex replied.

Terra knew that she was somehow seeing and sensing everything that Alex was seeing and feeling.

How could this be? she thought, almost aloud.

Her focus seemed to diminish as she heard the thoughts of others intruding upon her ability to follow Alex's thoughts. She cried out in pain and fear as she became overwhelmed by a flood of information. Then everything stopped, and she was alone with her thoughts. Whimpering, she ran into her parents' room. Her mother was up, sitting on the side of the bed.

"What's wrong, honey?" Riley asked.

"I don't know!" Terra broke down into a river of tears.

Riley hugged and held her daughter until Terra's sobbing slowed and eventually stopped.

"I could 'see' the ComCent room, Mommy," Terra said.

"What do you mean?" Riley asked.

"I was looking at the room through someone else's eyes!"

"Darling, you must have been having a lucid dream," Riley calmly responded.

"No, Mommy, I saw as she crawled through a shaft to get away," Terra explained.

Riley felt a coldness inside of her beginning to grow as she feared for her daughter. "What do you mean 'get away'?"

"There was no one else in ComCent, Mommy. She wasn't supposed to be there with the alarm going," Terra explained further.

"Who is 'she,' Terra?"

"Her name is Alex. She ended up talking to Gunner, Daddy, and the captain."

Riley looked at her daughter, perplexed by what she was hearing.

"It wasn't a dream, Mommy. It was real!" Terra stated defiantly.

Riley let go of her daughter, stumbling a couple of steps backward as she did. Neither of them said anything for several moments. Both stared at the floor, churning through their thoughts and emotions.

"Let's talk about this with Daddy in the morning when he finishes his shift. You can sleep with me tonight."

Both lay down and covered themselves with the sheet and blanket, remaining motionless for a long time before falling back asleep.

Toward 6:30 a.m., the wall slowly began lighting up, providing what amounted to a sunrise. By 7:00 a.m., Riley and Terra were awake, as Sam opened the door and came in, finding his wife and daughter in bed together.

"Sam, honey, Terra and I will be getting up. You can have the bed. What caused the pressure leak in the engineering spaces last night?"

"Nothing, there was no leak. It was a false alarm. Environmental is still trying to figure out what caused the alarm. It might have something to do with a failed ventilation shaft sensor. The ventilation shaft airtight door didn't close as it should have," Sam said.

"Daddy, that's the shaft Alex crawled through to get out of ComCent," Terra said in a matter-of-fact manner.

Sam stared at his daughter, silent and confused.

"Terra, was this a dream?"

"No, Daddy, I saw what Alex was doing. I saw you, the captain, and Gunner talking to her in the hallway. Gunner asked her where she had been. She said she went to the kitchen to get some soup."

Sam was now fully engaged in what his daughter was saying. He recalled the conversation between Gunner and Alex, trying to remember if somehow his daughter had been standing with the group of people in the hallway, desperate to find an explanation for what she appeared to know. There was no way Terra would have been awake and standing in the hallway with the throng of people who had gathered with the captain. Sam paused, carefully thinking about what to say next.

"Terra, why was Alex in the ventilation shaft?"

"She was doing something she should not have been doing. She was scared. She thought she would be caught."

"Terra, what was she doing in ComCent?"

"She was trying to look at a file that she was not supposed to look at. She was trying different passwords, but they didn't work, then she got scared and gave up and hurried out of the room."

A realization dawned on Sam that Alex had been trying to listen to or download the message from The

Mother. Sam knew that he had not downloaded the message and deleted it off the server yet, as the captain had asked Gunner and him to do. He resolved that he would get Gunner, go back to ComCent, and do it right away before he went to sleep for the "night." He had no idea how his daughter knew all this, but she did somehow.

Riley and Terra got up and went to eat something, leaving Sam sitting on the side of the bed. He thought back to the night The Mother appeared to him, Terra, Riley, and Gunner. He recalled his surprise when Terra knew what was going on before he told her.

Was this another instance of Terra knowing the thoughts of other people?

Regardless, Sam needed to download the message, delete it from the server, and secure it in the captain's safe. He knew he also had a serious security issue to discuss with Gunner and the captain. How could he explain this to the captain? Gunner was not a problem; he had seen The Mother. The captain had not.

Just then, Terra came back into the room and said, "Daddy, I will go with you when you tell the captain. I know how to explain this to her. Let me do it."

Sam reeled with confusion, as it seemed his daughter was reading his thoughts! He got up to get Gunner and headed back to ComCent. There was work left to do before he could sleep although it seemed that sleepiness had left him, leaving him buzzed and wide awake.

CHAPTER 23

To Catch a Spy

SAM, TERRA, AND Gunner sat in the captain's conference room, awaiting the captain's arrival. Gunner was agitated and expressed as much to Sam.

"We have no game plan! How do we explain all this to the captain?"

"Terra will explain this to the captain," Sam shot back. "She asked me explicitly to let her do it alone!"

For her part, Terra sat quietly, unfazed by the discussion that was about to take place. Sam had told Gunner about Terra's experience inside Alex's head on the night of the false alarm. Gunner had immediately gone into "threat mode." He had installed listening and recording devices in Alex's quarters and was now following her everywhere she went. So far, they had found nothing out of the ordinary, except that she was sleeping with Binna and another woman.

They all stood as the captain entered the room as custom dictated. As the captain sat down, they all sat down. An uncomfortable pause ensued before Sam said, "Captain, we have a grave issue to discuss. Terra will tell you about it."

The captain looked confused. “Sam, why is your daughter sitting here with us, and why will she be leading the discussion? She is, what, ten years old?”

Sam replied, “I understand your concern, but please, this is what *Terra* wants to do.”

“OK, little lady, you’re on deck,” the captain said dryly, looking skeptical.

Terra stood and began humming; then she started singing in an unknown language, all the while looking up at the ceiling. The captain and the others exchanged bewildered looks with one another, but then those looks melted away into astonishment as The Mother appeared in the room. The captain rolled her chair back, clearly frightened, but also confused as Sam and Gunner were smiling.

“What the hell, Sam!” the captain softly said, as she tried to stand and press her communicator to summon security.

“Sandra, you need not be afraid,” The Mother said in the same strange tongue, which was somehow understandable to everyone in the room.

Captain Simmons was surprised that her body seemed to take the statement seriously as her eyes returned to a normal size, her muscles visibly loosened, and her posture took a more relaxed position.

“I am known to Terra, Sam, and Gunner as ‘The Mother.’ I am your ancestor. I once lived on your world, long ago, before your history was ever even recorded. I now live on a planet in what you call the Cervantes star system, your new destination. Many of the women and a few of the men of my people can communicate telepathically and know the thoughts of others. This has led to us becoming a completely peaceful people. This gift will be returned to your people through Terra. She is now developing this ability. She knows of a threat to your mission because of this nascent ability. Pay heed to what she will now tell you," The Mother said. As she said this, she disappeared.

The room was quiet, as they all recovered from another otherworldly experience.

After a moment, Terra spoke. "Alex, in security, tried to download the message during the alarm last night. She did not get to it. She is working with people from Earth who are trying to stop our mission to find a new world. I know this because I saw through her eyes as she tried to do this. My ability to 'listen' to the thoughts of others is not sharp enough to control when I can do this. It just happens sometimes. I do not know why Alex is doing this. The Mother cannot listen in to the thoughts of people on our ship because of the huge distance between us and her. She has told me I am to use my new ability to help us in our mission. Also, she does not want me to go into hibernation during our voyage. She assured me I would not grow old, even after being awake the whole time. I am to let you know what Alex did so that you can decide what to do about it."

The captain spoke shakily. "Gunner, Sam, so you knew of this 'Mother' and didn't tell me?"

"Yes, that is correct, captain," Gunner said as Sam nodded in agreement.

"Why the hell did you not tell me?" she asked defiantly.

"Because she asked us not to tell anyone else until the time was right," Sam said.

The captain was silent for a moment then asked, "So . . . do we take Alex into custody, or do we watch her and try to learn more about this plot? Did she, or who she is working with, have something to do with the near miss we encountered a while back?"

"I don't think we know for certain, captain, but we could assume so for the moment, to be safe," Sam said. Gunner nodded in agreement.

"OK, let us keep an eye on her. Keep me apprised of everything she does. Gunner, you oversee this surveillance," the captain said.

"Already on it, captain," Gunner responded. "I'm listening to everything she says and watching everything she does."

"Good. Now, as to you, Terra, I really want to know more from you about why you have this special relationship with this Mother," the captain demanded.

"I'll tell you what she allows me to tell you, ma'am," Terra replied.

Captain Simmons looked offended, but Terra's firm expression made her realize she might not be dealing with the typical ten-year-old girl.

"OK, young lady. Come by my quarters with your father after your school classes end. We'll talk more then. I have a thousand things to do now, though none of them seem as interesting as all this stuff!" She waved her hand in a display of frustration as she said it.

CHAPTER 24

Invitations and Threats

New York City
Present Day

CARL PETERSEN SAT in his chair, staring at the New York City skyline, deep in thought and oblivious to the fact that Pritchard had entered his office. The building he was in was the old UN building. It had been renovated recently for a third time and was no longer the seat of world government; it was now just a rented space for businesses.

"Mr. Petersen, we have received a message from Alex. Our attempt failed."

"I figured as much. There was almost no way we could sneak up on the rock and take it out," Carl said flatly. "We must think outside the box, Pritchard my boy. Get the seven together tonight. We need options, good ones."

"Yes, sir," Pritchard said, as he turned and walked out.

"The Seven" referred to six business executives and one MIT physicist, Dr. Ted Bywater, who worked under Carl. Each of them was a greedy, sociopathic, ambitious bastard, in other words, a good CEO who never failed to miss an opportunity to make money and amass more power and influence.

Bywater was the one who had received the first message from "the Association of Worlds," or the AOW, as they referred to it. The invitation had come via the "gadget," which had been found on the island of Crete in an archeological dig, one hundred years earlier. No one had the faintest clue as to what it was until Bywater had theorized that it might be a communication device left by ancient space travelers over fifty thousand years earlier. He ended up being right. The device, he theorized, contained a "zoo" of entangled particles that could be used to communicate with whoever held the other half of the pairs of entangled particles. Because of this, Bywater was awarded a patent on the most significant advance in communications ever. The *Calypso* utilized this technology to communicate with Earth.

Bywater wanted more. He wanted to talk to whoever created the gadget. Surely, they would have additional technology that could make Carl and him even more wealthy. To talk to the "little green men" required that they know something about their language. *What language would the aliens understand? Mathematics?*

Bywater speculated that whoever left the device was listening to our broadcasts that had been expanding into space since the first days of radio and TV centuries earlier. He believed they would know how to interpret most of the languages on Earth based on these broadcasts. He was correct again.

The device had a screen with which they could draw letters and symbols. In a move of pure desperation, they wrote a message in modern English and pushed a small button on the device; and by God, they got an answer moments later—in English! This led to a steady stream of messages back and forth that ended with "the invitation." Really, it was more like a contract.

Help us and we will give you technology that will make you rich and powerful gods on Earth.

Who could pass on such an offer? The "help us" part was not elaborated on; they were just to do as they were told, and they had been told to destroy the *Calypso* by any means necessary.

As time passed, the messages began to take a sharper and more menacing tone. "They" wanted status reports. They wanted results, and now there were threats—*get this done or your species will be wiped out.* Carl was afraid of the box he had opened. The thrill of communicating with extraterrestrials was gone, and the numbing coldness of fear in his gut had replaced the thrill. Carl truly believed these were beings who could make good on their threats. He did not want things to get to that point. After all, his life was at risk as well as his family, and his fortune most of all.

CHAPTER 25

Make Them Glow

THE FACILITY NUCLEAR reactor was located one hundred meters below the living and engineering spaces. The output of this reactor could power a medium-sized city. In fact, the needs of the *Calypso* were equivalent to a medium-sized city. The power requirements of oxygen extraction from carbon dioxide alone were enormous. Add to that the need to run the water reclamation plant and the station power requirements, and you had a demand for an impressive amount of electricity. Because of this, nuclear power was the only option for the mission. There were battery and solar power backups, but they could provide only enough power for the environmental systems, the systems that kept people warm and breathing, and then it was only good for ten to fourteen days. It was assumed that any problem that shut down the reactor could be fixed in that time. That was the thinking of the best minds who had designed the *Calypso.*

Alex had received a message on her entangled particle comm device the previous day. The message was simple—destroy the facility reactor water pumps and breach the reactor coolant water reservoir. A double blow would ensure the death of everyone on board. Granted their deaths would be slow and agonizing, but she had been picked because of her inability to feel the pain of others. Alex was a classic type 1 sociopath. She was particularly good at faking her emotions on her emotional psych evaluation, which she passed with flying colors. For her, the only thing in the universe that mattered was, well, her. Her mission was never meant to be a suicide mission. She would eject herself on an escape pod and put herself in long-term hibernation. The AOW had assured her she would be rescued, perhaps many decades in the future, but she would be a rich woman with a bright personal future.

She needed to access the facility nuclear reactor. This required some thought, quite a bit of thought. She would have to produce a badge and some biometric material, DNA, or even somebody's finger . . . ouch! It was also a bit concerning to her that she was running into Gunner all the time now. She wondered if she was being watched, or if he suspected something. In the end, she decided she had to ignore those worries. Entangled particle communication produced no signals that could be intercepted, and there was no way to decrypt them since no one had the other half of the particles. So she thought she had nothing to worry about. Except she still was not 100 percent certain that she was in the clear. She hated being a worrywart, but it had kept her alive in some tough spots in the past.

Alex walked briskly down the hall to the first of two hatchways leading into the reactor space. At the first one, she pressed the ring button.

"State your name and purpose," came the response on the video monitor.

"This is Alex Cornelius, deputy head of security, here for an inspection of the surveillance systems," Alex answered.

"We were not notified of an inspection by Gunner Lawson, your superior. He will need to contact us, ma'am," came the reply.

"Let me get in touch with him and get back to you, hopefully today," she answered, frustrated.

Gunner was up to something. *Why would he assert control over the inspection of surveillance systems?*

She had a plan B, and it involved Glen Lessard, the lead reactor engineer. Glen was single and had tried striking up a relationship with Alex on more than one occasion. He did not have a clue as to her preferences.

Clueless dick, Alex thought. *This should be a piece of cake!* She would seduce him and use him to gain entrance to the reactor space.

That evening, Alex made her way over to Glen's quarters. Glen was happy to see her. They stripped their clothes off and began their lovemaking. As Glen finished his thrusting, Alex climaxed, arched her back, and moaned loudly. The two of them were breathing heavily and sweating. Glen rolled over. They were both on their backs now, staring at the ceiling, silent.

"Alex, you're a wildcat!" Glen said.

"Glen, you sure pack some heat!"

They both faced each other.

"We'll have to do this again soon," Glen added, still breathing heavily.

"I would like that. Tomorrow afternoon? I'm hoping to be done with the surveillance inspections by then. Your area is my last one to inspect. I tried to do the inspection in your area earlier today but got told Gunner needed to approve it. Anyway, I can be done in a few minutes, but if I must chase him down for permission, it could crowd out our next rendezvous. Can you get me in right after lunch, say, 1:00 p.m.?" Alex asked.

"Should be able to. Gunner likes to play like he is a monarch sometimes, lording over small stuff like 'permissions' and such. His Majesty can be a pain in the butt sometimes," Glen added. They both laughed at that.

"Well, I better get going. See you tomorrow," Alex said.

"See you later, wildcat!"

Later the next day, Alex stood on a stepladder in the reactor control room, adjusting camera views, as Glen came over to her, gave a wink, and asked, "Just about done there?"

"Almost. I just need to see if it gives a view of the control panel over there," Alex said. She moved over to the control panel, all the while looking at the view from the camera projected onto her tablet. Just then, one of the reactor techs came up to Glen and started a conversation about the reactor room coolant pumps. Alex saw her opening and interrupted the technician.

"Glen, we don't have a security camera in the reactor room. I know we've talked about this before, but we really need one," Alex said.

Glen nodded and told the technician, "Steve, take Alex into the reactor room when you go back in and help her set up another camera, can you?"

Steve answered, "Sure, I'm going back in right now."

In the reactor room, Alex had Steve hold a ladder while she climbed it and mounted a new camera.

"Steve, I want to check the view from the camera. I left the screws loose so it can be adjusted and then tightened when we get the view I want. Can you climb on the ladder and move the camera as I tell you?"

"Sure, hoss boss," Steve answered.

Alex went to the guardrails encircling the top of the reactor and began looking at her tablet and the camera view. She talked Steve through some adjustments. As he made the adjustments, she attached a high-explosive block next to the main coolant line that headed into the reactor. The explosive had a very sticky residue that held it in place and a small timer and detonator. She had set the timer for ninety minutes. All that remained was for her to get out of the reactor room and wait for the explosion while she readied an escape pod. She had put another explosive under the main control panel and had set it to go off first, three minutes before the explosive on the coolant line went off. With this completed, she went back into the control room and made a quick excuse to get her out of her “appointment” with Glen.

“Glen, I’ve been summoned by His Majesty Gunner for a security review. I’ll need a rain check,” she said.

Glen looked disappointed and muttered, “Sure, let’s talk later.”

CHAPTER 26

Emergency!

GUNNER WAS RUNNING toward the security office where he had asked Alex to meet him. He had lost track of Alex's movements and needed to reacquire his target ASAP. This was the first time he didn't know where Alex was in the days since he began following her, and he had a sick feeling that something bad was going to happen.

When he arrived at the security office, he looked for Alex but did not see her. He cursed and called the captain to tell her about the situation.

"Captain, Gunner here. I've lost Alex. She gave me the slip. This is the first time. I have a bad feeling. Recommend we initiate a security lockdown," Gunner said, out of breath.

A security lockdown meant that everyone was to assemble at their lockdown location and be accounted for by their supervision.

Ten minutes later, all awake station personnel were at their designated rally locations. Alex was to assemble with Gunner in security, but she was nowhere to be found.

Gunner called the captain. "She is not here. We need to assume she has done something that will harm the ship and personnel. Captain, recommendations?"

"Gunner, get Sam, Riley, and Terra. Go to my cabin. I have an idea!" the captain said with urgency.

As they gathered in her stateroom, Captain Simmons explained that Alex was not accounted for, and she may have been planning an act of sabotage. "Terra, can you 'see' Alex? What is she thinking?"

Terra exclaimed in terror, "I see her near some little ships, pods? She is trying to escape something that will happen in minutes! A bomb!"

"Terra, where is the bomb? Do you know what she has done?"

"In the reactor room and the room with consoles."

"Can you see the bombs?" the captain asked, pressing Terra for more details.

"One is on a big pipe. It says, 'Coolant Water.' Another is under a console," Terra blurted out, fear on her face.

"Oh my god! The reactor coolant! She is trying to cause a pressure buildup and explosion. Get the reactor team to search for the bombs. Gunner, head down there!" Simmons shouted to Gunner.

There were two other security personnel with them who had gathered for the security lockdown. Simmons said to them, "You two, follow us to the pod bay!" Continuing, she said, "Sam, Terra, Riley, you three come with me to the pod bay also. I want Terra to try to see who Alex is working for. Perhaps there are others on board."

Once in the pod bay, they looked for Alex, but she was nowhere to be found.

"Maybe she is hiding in a pod?" Sam asked.

"She's already left," Terra said.

As they scanned the rows of escape pods, they noticed an empty spot in the front row. Captain Simmons tapped her wrist communicator and spoke. "Peter, get the shuttle fired up, we need to find Alex's escape pod. It couldn't have gone far yet. Get topside cameras up and look for the pod visually. Wait, what am I thinking?" She tapped her communicator and spoke to Carl Carroll. "Carl, get your people looking at the radar for Alex's pod!"

"Copy that, captain. Manu is in the observatory control room. He'll be looking for it."

"Captain, I have a lock on her pod. Do you want me to take her out with the rail guns?" Manu asked breathlessly.

"Negative. Hail her, get her to land topside!" Captain Simmons shouted back.

"Captain, she is not responding," Manu responded. "She's maneuvered to our aft . . . captain, she's heading toward the nozzle assembly!"

"Take her out, Manu! Take her out now!" the captain screamed.

"Copy that!"

Manu authorized the SDS to destroy the target, which it did with one pulsed round from the rail gun.

"Captain, Manu. Target destroyed."

Meanwhile, in the reactor spaces, Gunner and the reactor personnel had located the high explosives and shut down the reactor for safety purposes.

"Captain, what should we do with these explosives?" Gunner asked.

"Eject them in a pod," Peter interrupted.

"Can you tell where these explosives came from?" the captain asked.

"They're from our own stores, captain," Gunner answered.

"I want you to search her quarters, interview anyone with whom she was close. Terra, I want you and Sam to be present when we interview those who knew Alex. I need to know if we have more of the enemy among us."

"Captain, what is this?" a security officer asked, holding out a device with a keyboard and screen, attached to a cylindrical tube.

"It doesn't look like an explosive. Get it to the spent fuel vault, just to be safe. Peter, Stuart, meet me in the vault. We have a gadget of some kind. I want to know what it is. Sam, come with me to the vault as well, I have an idea what this is, and it may fall into your area of expertise."

CHAPTER 27

A Deliberate Deception

THE GADGET WAS set on a small table in the vault. Peter, Stuart, Sam, Gunner, and the captain all stared at the item.

"Could this be an entangled particle comm device?" the captain asked.

There was a babble of conversation as each one gave an opinion, with the consensus being that it was an EPC device although it was different from the designs with which they were familiar.

The question was answered definitively a moment later, when a message scrolled across the screen that read, "Status on *Calypso*? Reactor overpressure achieved? Require urgent answer before you hibernate."

They all stared at the device in silent amazement.

Peter eventually spoke, carefully choosing his words. "Captain, if we answer that the 'mission was successful' or some such phrase, whoever is on the other end will think we are destroyed. Seems like that would take the target off our back."

Sam spoke next. "We do not know who is on the other end. What if it is in Mission Control? They will think we have been destroyed. Do we want to do that?"

"Right now, we do not know who is working for us and against us. We can do this mission on our own. The mission was designed that way. What we do not know is who else on board is working against us. What is the status of the interviews with Alex's acquaintances?" the captain asked.

"We have talked to Glen from the reactor control room, and Binna and Stephanie Pokorny, two women who have been sleeping with Alex. Terra says they all knew nothing about Alex's activities. There are others we are still talking to, but I think she was acting alone. She was trying to escape by herself so . . ." Gunner trailed.

"I agree, she was probably working by herself on the ship," the captain responded.

They discussed how to answer the question and agreed that they should say the mission was successful and the *Calypso* was destroyed.

"What about any code words or security double-checks? We might not answer in the correct manner. It could give us away," Sam said.

"Well, there didn't appear to be anything other than a straightforward question on their part, so I think we answer the same way," the captain said. They all agreed.

The captain typed, *Mission success.* Calypso *destroyed.* She pressed Enter, and the screen flashed, *Message sent.* In less than five seconds, an acknowledgment came back: *Splendid, well done. Proceed to hibernation and await pickup.*

Sam asked, "The next scheduled EPC to Earth is this evening at 7:00 p.m., just one hour from now. Do we just go 'dark' on Mission Control?"

The captain thought for a moment then said, "We do, at least until we figure out who is on our side."

Peter looked unconvinced. "Look, I understand why we should do this, but this will cause all sorts of issues back home, captain."

"I'm aware of that, Peter. Even so, our best move is to be silent. I want a blackout on all light-speed comms also. We need to be 'dead' to everyone."

"We'll need to tell everyone something. They are expecting communication with loved ones," Stuart said.

The captain thought for a moment then said, "Tell them we have been ordered to go radio silent for security reasons. That is more or less truthful anyway."

"Copy that," they all said.

"One more thing, captain. We need to hide ourselves from visual observation. Recommend a course adjustment. We should move ourselves off course, and we should shut the ion drive down for a few days. We should do it right away before they try to see us from their space-based telescopes," Peter said.

"Make it so. Stuart, Peter, make an announcement that we will be going weightless for a couple of days without the drive accelerating us. Say it is to do maintenance on the drive system or something like that," the captain ordered.

"Yeehaw! Zero g's!" Gunner whooped.

CHAPTER 28

Mission Control, the Earth

KWAME OLAJUWON, *CALYPSO*'S mission manager, had a problem. The *Calypso* had failed to respond to their evening comm check. It was midnight now, seven hours after their missed check-in. Besides missing one of their twice-daily "chats," the telemetry data was also missing on the *Calypso*. He had ordered that the Tyson space-based observatory do a visual track on the *Calypso*. Normally, he would have waited, but certain recent events on Earth had caused his anxiety level to increase to ten out of ten.

The FBI had raided the home and office of a trillionaire, Carl Peterson, as part of an investigation into insider trading of stocks. When they looked through his electronic communication, they had found communication between him and five other wealthy individuals and the inventor of entangled particle communication, Dr. Ted Bywater, that seemed to indicate a plot to destroy the *Calypso*. Kwame had been notified the day before of this, and now the *Calypso* was not responding. Kwame was worried, and rightly so, he believed. He had kept the information about the possible plot to himself, as the FBI had asked him to keep the information secret so they could investigate without tipping off any coconspirators.

He needed to know what was happening on the *Calypso*. He decided he needed help not only from the FBI but also from his most trusted adviser, Nigel Smythe, a one-hundred-twenty-year-old engineer who had led the space agency for over sixty years and who was now retired. Nigel was his mentor and was always available to help Kwame, whom he treated like a son. Kwame decided he would have to do what all good managers do, but which he hated—call a meeting for later that day.

"OK, so I need to tell you something, Nigel, something confidential," Kwame began. "The *Calypso* did not return comms last evening, nor this morning. This is the first time in almost a year that the comm cycle has been broken. Additionally, our space-based telescope couldn't locate the nozzle plume. They are still close enough for us to see it, to answer your question in advance. Two days ago—and I know Agent Ferris knows this—the FBI notified me of a plot to destroy the *Calypso*. Six trillionaires and one MIT physicist, Ted Bywater, were arrested in an insider trading investigation. When they reviewed the evidence, they found out there were discussions of destroying the *Calypso* and payments of some kind for doing that. I have a feeling these events might be related somehow. I do not know how, but before I announced to the world that our mission was lost, I wanted to consider other prudent courses of action. Thoughts?" Kwame asked.

"Couldn't the mission have been lost through one of a million different technical things? Isn't that the much more probable cause?" Nigel asked.

"Well, yes, but we live in a different world now. Very little of the technology we chose for the *Calypso* was cutting-edge. We have been traveling in space for hundreds of years. There have not been any losses of spacecraft for more than fifty years now. We know what we are doing. Besides," Kwame said, "I can live with a technical cause. I cannot live without taking this threat seriously, and that is why I want to think before I act, but I want help with this."

Agent Ferris spoke up. "I agree. Consider if you have an insider threat and someone in your agency is privy to comms with the *Calypso*, they know right now that they are not responding. They may be reporting that the mission, the plot, was successful. This means whoever is behind these efforts will believe they succeeded in destroying the *Calypso*. We may want them to believe that. Consider also that perhaps the *Calypso* found out about the plot and prevented it. Or, they are damaged, but still alive. In these cases, they went dark because it was in their best interest to do so. Who knows? But play this one card. Before you go public about the loss of the *Calypso*, try a back channel to them if you have one. Kwame, Nigel, does your agency have any other roundabout way of communicating with the *Calypso*, a way that might not be detected by anyone working in your mission group?"

Kwame and Nigel sat and looked at each other for a moment, saying nothing.

Finally, Nigel spoke. "Ah yes, there might be.

"Kwame, you heard, or read, or maybe not, about the loss of the *Locator*, oh, fifty-five years ago now, I believe? The *Locator* was an asteroid mining vessel that had an engine failure. The early ion engines were prone to overheating. Well, the entire crew was lost, no way to get them, so they died, slowly, drifting in the asteroid belt. That ship can still pump out a message. We know—we went back to it thirty or so years ago. A documentary crew wanted to do a story on the ship, and they decided to go out to her and board it. Our agency insisted that the mission be under our command. When we got there, we found that the ship designers had installed backup squib-activated batteries for comms and life support, which had never been activated. I know, old technology, but very, very reliable. The electrolyte is not mixed until it is squibbed, so the batteries have an exceptionally long storage life. The ship got sold and the new owners never got all the technical manuals. They never knew about the battery backups, so the crew that was lost never knew either. I do not know if it would have made a difference in terms of their rescue, but a few weeks of life support and comms were available on battery backups. We could command the activation of the batteries then send a tight beam message meant for the *Calypso* to the *Locator.* The *Locator* would wait for twelve hours and then send our message in a tight beam to where we think the *Calypso* is located. They would get the message if they were still operational. They could see where it came from, a derelict ship. That might convince them you are playing it safe for their sakes, that you are the mission director, and that you are on their side, not a plotter. Maybe they would talk back to us with that knowledge. Possibly. That would be done via light-speed comms, not EPC, but the plotters probably don't have the ability to intercept light-speed comms anyway. At any rate, the bad guys would have no way of intercepting this type of communication coming from a dead ship. They simply would not be expecting it. They would not be listening for this."

Kwame and Agent Ferris sat in silence and stared at each other.

Kwame began. "OK, yes, let us assume they found out about the plot, foiled it, but went dark to protect themselves from somebody here at our agency. This seems unlikely, but I see no downside to trying a back-channel comm method, one they feel safe about. We'll try it. Agent Ferris, any objections?"

"No, none," he answered.

CHAPTER 29

Answer the Phone!

"CAPTAIN, THIS IS Sam. I think you should come down to ComCent. I have a light-speed message from an unusual source."

When the captain arrived, Sam played the video message:

> *Calypso*, this is Kwame Olajuwon. A few days ago, several trillionaires, led by Carl Petersen and physicist Ted Bywater, were arrested for financial crimes. The FBI found discussions of a plot to destroy the *Calypso*. Right now, we do not know much more than this. The next day, you, or the *Calypso*, stopped answering comms and we couldn't locate you visibly. Only myself, my mentor, the retired director Nigel Smythe, and one FBI agent know about me sending you a 'stealth' message via this circuitous method. We did it because we didn't want to alert any potential plotter who may be within our organization. If you are able, please

> respond with your status to the old *Locator* at the coordinates supplied via tight beam. We are working to secure comms by EPC by carefully vetting our personnel who will be operating those comms. Until then, we will be communicating through light-speed comms on this message routing. We've told everyone that there was a comms issue, and we don't know your status. We know we took a risk broadcasting to your comms in the event you may have personnel who you don't trust, but we wanted to take that risk. I'm really hoping you folks are OK.

The captain and Sam were silent for a moment; then Sam asked, "What do we do? Do we respond?"

"We respond, give them our status, tell them about Alex, and let them know about the communications we've had with The Mother. We need them to know what we know. Maybe all these unusual events are in some way tied together. But we need to reestablish comms again in a secure manner. This stays between you, Gunner, and me for now. I'll draft the response. We want this to be a video. I want them to see and hear me only," Captain Simmons said.

"Copy that."

The *Calypso*'s ion drive had been reengaged after two days. They knew they would have a course adjustment because of the lost time coasting rather than accelerating. They would need to recalculate their flight profile, but the change would be minor, they hoped. The guidance engineers were working it out. It might not be fully optimized like a change developed by Mission Control, but they would make do. Many years later, there might be more course tweaks to account for the less than 100 percent precision.

Later the same day, the *Calypso* had a return ping from the *Locator*, indicating it had received their return message. For now, they had to wait and see what Mission Control could do with vetting personnel who were on the Calypso ground control team.

CHAPTER 30

Metropolitan Correctional Center (MCC), Lower Manhattan

"MR. PETERSEN, I'M Selma Richards-Gomez from the US Attorney's Office for the Southern District of New York. Also with me is Seth Williams, also from the same office. We have previously offered a plea bargain proposal to you and your counsel, Mr. Goldstein. I am here to hear your answer to the plea deal."

"Counsel, my client has agreed to your plea proposal in full. He will cooperate against those who offered payments for the plot to destroy the *Calypso*," Goldstein answered.

"OK, Mr. Petersen. One, how did you plan to destroy the *Calypso?* Do you have a saboteur on board? Two, who offered compensation to you for this action?" Selma asked.

Carl answered. "We had a person on board the *Calypso*, Alex Cornelius, the deputy security officer who planted explosives in the facility nuclear reactor. This disabled the coolant water piping and caused the reactor to overheat and melt down. In turn, this destroyed the integrity of the pressurized living compartments, killing everyone on board."

"Mr. Petersen, are you saying that your plot succeeded and did, in fact, destroy the *Calypso*?" Selma pressed further. "How do you know this? Are you in contact with this individual?"

"We are. We received confirmation."

"How did you receive confirmation? Aren't Mission Control the only people who can communicate with the Calypso?"

"Alex had an entangled particle communication device. She sent a message that she had destroyed the ship. We received it two days ago."

"I am not aware that the ship has been destroyed. Are you sure of this, Mr. Petersen?"

"Yes, the ship has been destroyed, Counselor," Carl answered.

"I am going to seek confirmation on this from the space agency. Please continue."

Carl continued. "The entangled particle communicators in use today on advanced spacecraft are based on an artifact discovered in an archeological dig over one hundred years earlier on the island of Crete. Dr. Ted Bywater, who owns the original patents on entangled particle communication, had this device stolen from the Smithsonian. He figured out how the device worked—that is how he came up with the patents he received. He fabricated a pair of devices, one of which was given to Alex Cornelius. This is how we received confirmation of the destruction of the *Calypso*. Alex Cornelius sent a message to us via her device after she escaped the *Calypso* in an escape pod."

"Mr. Petersen, are you saying that an item from antiquity, thousands of years ago, is an entangled particle communicator? I find this impossible to believe."

"Yes, that is what I am saying, and this leads us to who offered to pay us for this action. The artifact found on Crete was left, or lost, by technologically advanced aliens, thousands, perhaps tens of thousands of years ago. That same race of aliens responded to a message we sent to them on that device several years ago. We have had numerous conversations with these aliens. They knew of the construction of the *Calypso* and its mission. They were very earnest in their desire to have the vessel destroyed and sought our assistance in doing so. They offered us advanced technology in exchange. We agreed to do it. We felt it was worth the cost for humanity to have their technology, or at least some of it."

"Are you saying you destroyed the *Calypso* solely for the benefit their technology would bring humanity, Mr. Petersen?" Selma asked.

"Well, it would also bring me and my colleagues financial gain," he answered sheepishly.

"I see, Mr. Petersen. What do you know of these aliens? Are they a threat in any way to our planet?" she pressed.

"Ah well, they may be. OK, they are a threat, Ms. Richards-Gomez. They have threatened me and my colleagues. They have threatened to wipe out our species if we fail. This is my motivation for accepting your plea offer and divulging these details."

"Mr. Petersen," Selma asked, "what else do you know about these aliens?"

"They said they represented the 'Association of Worlds,' or AOW."

"Where is their planet located?" she asked.

"They would not answer that question. They refused to answer our questions, questions that many people would want to ask aliens if they appeared before us on Earth."

"I see, Mr. Petersen," Selma responded.

Carl continued. "I believe our actions have spared our planet from destruction."

"I hope you have, Mr. Petersen. I hope you have. Neither you nor your counsel are to discuss this with any other people. I need to discuss this with my superiors and with the space agency."

"Do you believe any of that nonsense, Selma?" Seth asked after they left the interrogation room.

"I don't know, but I'm not ruling anything out. We live in strange times. We've got work to do," she continued as they walked out of the building.

CHAPTER 31

Deceiving the Enemy

"DIRECTOR OLAJUWON, THANK you for meeting with us," Selma began. "Agent Ferris has briefed me, the secretary of defense, and his primary assistant, Willie Turner, on the communication you have reestablished with the *Calypso*, the activities of Alex Cornelius, and the communications you have had with the alien that is being referred to as 'The Mother.' We are gratified that the crew is safe and the mission is progressing. Agent Ferris has told me about your concern about an insider threat that may remain within your mission team. We have had members of your team under surveillance for a little over two weeks now without any hint of nefarious activities or discussions. We wanted to brief you on an extraordinary threat that our world faces that we have learned about from our discussions with the suspects, Carl Petersen and the others."

She paused then turned to Willie Turner, the assistant secretary of defense. "Secretary Turner is here today to discuss how we will handle this threat."

"Thank you, Selma. The Department of Defense has a keen interest in all of this, given the obvious national security implications. Carl Petersen and his cohorts have also been in communication with aliens via an EPC device discovered on the island of Crete that dates back tens of thousands of years. He was offered advanced alien technology in return for destroying the *Calypso*. The aliens he has been communicating with have made threats to our entire planet if he does not destroy the *Calypso*. For the time being, they, the aliens, have been led to believe Carl and his cohorts did, in fact, destroy the *Calypso*. We want to keep them thinking that for obvious reasons. They represent some sort of organization that they referred to as 'the Association of Worlds' or AOW. Mr. Petersen has learned almost nothing else about these aliens—they refuse to share any useful information with him."

Turner paused then looked directly at Kwame and continued. "I want the public to be told that the *Calypso* was destroyed due to a reactor malfunction. We do not want the AOW to learn that the ship and mission are intact. Because the crew has been in communication with what we believe is another group of aliens who are friendly, we want the mission to continue. We may need allies, allies with advanced technology. We do not know if we can successfully defend ourselves against the AOW with our technology. We believe we need to foster the relationship with the people of The Mother."

He paused again then continued. "I want your Mission Control team to be disbanded. We will establish a new mission team, headquartered at Groom Lake, Nevada.

You can handpick some individuals to remain on the team subject to successful background checks. Groom Lake is ideal for a covert mission like this because there are other ongoing classified development projects. Another classified group will not draw attention from the public or other folks in the military or scientific community. This project has been designated as 'Temple Mount' and has the highest classification level, top secret, SCI, presidential-approved access only. Because Mr. Petersen and the physicist Dr. Ted Bywater have familiarity with these aliens, I want them on your team. Subject to their plea agreement with the SDNY, they are to offer full cooperation in exchange for avoiding prison time. We will always have them under surveillance. For now, until we have the proper people and equipment set up at Groom Lake, you should continue your communication with the *Calypso* via the *Locator*.

"We do, however, want to return to EPC as soon as possible since no one can intercept those comms. We expect this to be only a few weeks. You will need to fully brief Captain Simmons on all of this. She will determine who to share it with among her crew. The comms from the *Calypso* need to have the highest security so that no communications will ever be made except to you or members of your new team. You will announce to the public that the mission was lost due to a reactor meltdown. You will resign as head of the space agency. The public will believe that you took the blame for the mission failure and were forced to resign. Agent

Ferris and Ms. Richards-Gomez will be part of your new team and will be the primary liaison with myself and the president. I can't stress enough the need for total secrecy—all our lives may be at stake. We appreciate your cooperation."

CHAPTER 32

Solace for the Dead

CAPTAIN SIMMONS FINISHED briefing her crew. Everyone on board was affected by the events on Earth and on the ship. There had been questions from the crew as they assembled in the shuttle hangar bay; sometimes an answer was possible and given, but often, there was only conjecture. No one knew what the relationship was between The Mother and the AOW, for instance, except that the AOW might be the enemy to whom The Mother had referred. But the biggest concern was the loss of communication with loved ones on Earth. Since the ship and crew were declared dead to those on Earth, they obviously could not have any messages with loved ones. This weighed heavily on everyone. Even though they knew their trip was a one-way endeavor, they had expected to have some communication with loved ones; and now, unless something changed, they would not.

Captain Simmons believed she had the cooperation of all the crew. She believed they understood the vital need to maintain the cover story that the ship was lost and the crew dead for the sake of the safety of all the Earth, including loved ones they had left behind. She had stressed that their crewmates were their family now, and they should find solace and companionship among one another. She believed this would help the crew develop comradery since they were all in it together. Even so, there was a palpable feeling of loss among the crew as if a death in the family of every crew member had occurred. They had all been chosen primarily for their emotional resiliency. A one-way voyage was not for the fragile. Eventually, things would return to a semblance of normalcy.

Now that the entire crew knew everything, Terra found herself the subject of many conversations. Everyone wanted to know if she was reading their thoughts. Terra explained over and over that she had to consciously exert effort to see what others were thinking and it was emotionally exhausting to couple with the thoughts, emotions, fears, and memories of others. She did not want to do it unless it was required, but she also sensed that people were viewing her with a respect that puzzled her. She was, after all, only eleven years old. In so many ways, she wanted to remain a child and play with her friends, but she found that they no longer looked at her that way. Fame had left her alone and lonely. She needed a close friend, someone in whom she could confide. In time, she came to realize that Hatcher was such a friend. Terra liked that Hatcher did not treat her differently; instead, he remained playful, yet confident around her. She was, in fact, developing the friend that she needed.

"So what will we call that purple nebula?" Terra asked playfully.

"Cassia's fart!" Hatcher answered, laughing heartily.

Terra started to laugh then caught herself and answered, "Hatcher, that's not nice!" They both laughed together.

They were in the observatory with the astronomer on duty and were looking at the big screen showing the star field in front of the ship. They came here often, peppering the astronomers with questions about the stars in their view. When a star or a comet did not have a name, they would offer suggestions, usually silly ones. As time passed, given their velocity, star positions and other objects slowly changed, but the number of stars visible, even visually, was stunning. No atmosphere, no city lights, no clouds, no pollution: a perfect recipe for enhanced star viewing.

Hatcher would walk Terra back to her family's quarters each time they finished stargazing. Terra felt herself beginning to like Hatcher, and she knew he liked her. She thought at times that this was due to the lack of available kids, but she decided that wasn't the case. She really did like him.

CHAPTER 33

Strategy Sessions
Groom Lake, Nevada

THE EARLY MORNING air was cold, the sky a perfect blue with no wind. A typical January day at the dry lake. Kwame thought he might get in a round of golf in Vegas on the weekend, but for now, he had an important strategy session with his leadership team: two assistant secretaries of defense and their space force teams and six highly regarded academics who were experts on, among other things, "alien civilizations." They needed to try to anticipate things, things that never could have been anticipated. It would be a lot of SWAGs or "scientific wild-ass guesses."

Kwame stood outside at the far end of the parking lot. Soon, the hovercars began arriving from Nellis Space and Air Base at 9:00 a.m. The meeting would kick off at 10:00 a.m. in the large conference room in Building Ten. Because it was top secret SCI presidential-approved access only, the meeting required that every attendee be scanned for electronic or biometric devices prior to entering the room.

This will take some time, so it was good that military folks always arrived early, Kwame thought.

The large conference room was an anechoic chamber. It was completely protected from outside electronic emissions, and it would not allow any electronic emissions to escape the room. The thinking was that the room was a completely safe place to discuss the most classified of subjects. And it was, at least as far as the technology of any person on Earth. But it was not safe, it turned out, from technology beyond what humanity possessed.

It turned out the human mind was not just neurons, blood vessels, and other tissues. It had quantum aspects to it, ones that recorded every sensory input in perfect condition.

There were particles produced when neurons fired. They would pop into existence then pop out of existence as quickly as they appeared. These quantum particles would interact with other particles that would, in turn, interact with other particles and so on. Their mass was infinitesimally small, and they could pass directly through anything, including the Earth. Like a pebble dropped in a pond, the collisions of these particles produced quantum waves, which were completely independent of electromagnetic forces and thus could be "read" if one possessed the right technology, *or biology.* This effect diminished, however, with the square of the distance between the person whose mind one wanted to read and the receiving technology.

The Mother had this ability as part of her biology, and its range was large but not unlimited. The effective range of this AOW technology was something on the order of a few hundred miles, which meant Las Vegas was certainly in range of the ripples. But a person also needed to know the precise location of whoever they wanted to listen to. That had turned out to be easy enough. The military officers had gotten drunk the night before and had talked about their "top secret meeting" the next day.

In a hotel room where the old Trump Hotel had once been, a man, Effam, sat in front of several consoles, listening to the ripples produced by the thoughts of those in the classified room at Groom Lake. The thoughts were precursors to words that came from the mouths of the attendees, so the information gleaned was of the highest quality in terms of signal intelligence. After the meeting concluded, Effam pulled out an EPC device and typed in a message. The message alerted his superiors that the *Calypso* was not destroyed, that Terra still lived, and that she could still fall into the hands of their enemies, the people of The Mother. He awaited the response on what to do next, which came almost immediately:

Return to Lunar Base.

By the next day, Effam was on the moon. He sat on one side of the long table; his superior, Kaanon, sat on the other side along with three others. They were sixty meters below the surface on the plain of Hadley Rille. This was west of where the Apollo 15 lunar module had landed in the twentieth century.

"Petersen and his colleagues were not truthful with us," Kaanon said. "Or their person on the *Calypso* lied to us. Either way, you have failed in your mission. Failure is not tolerated in your position. Do you have anything to say?"

Effam shouted in desperation, "Master, I believe The Mother's people produced more than one telepath. Allow me to search for other telepaths. I beg of you, let me find others!"

"There are no second chances. You knew this. Take him to the air lock now!" Kaanon shouted.

Two of the three men seated with Kaanon grabbed the restrained Effam and hauled him from the room as he continued pleading for his life.

Kaanon sat silently then spoke to the remaining person, Esam. "He is right in assuming there must be other telepaths. I want you to find others and bring them here. We will not let her get possession of any more telepaths."

Esam spoke. "Master, our Lords wanted these telepaths taken to their world. Even if we found them today, we would not be able to get them to our world before the *Calypso* reaches The Mother's world. How will this help us?"

"We may have found a new way. Esam, find them as soon as possible!" Kaanon said.

"What do you mean, my master?"

"The speculation of passageways between locations in the universe has been confirmed. We have found one of them, a wrinkle. We may be able to get telepaths to our world much sooner than we are able to do now. Go, do as I have commanded!"

Kaanon realized that his life was in danger if he did not get telepaths. If he couldn't, he was to destroy the Earth to prevent The Mother from getting any of them. Kaanon didn't want that to happen. He still viewed the Earth as his home, and no one wanted to burn down their own house.

He stared at the wall and thought, *The AOW will still have to kill Terra, the telepath on the* Calypso. *Killing her is out of my hands now. My superiors will have to do that by coming through the wrinkle, intercepting the* Calypso, *and killing or capturing Terra.*

CHAPTER 34

Mission Status

EPC COMMS WITH the *Calypso* had been reestablished, and to a significant extent, things were restored to normal, except that the loved ones of the *Calypso* crew now thought they were dead. Kwame didn't like that at all; he could hardly imagine what the crew was going through.

Kwame didn't know it, but the underground bunker that now served as Mission Control was not known to the AOW. They couldn't "listen" to anything there because they didn't know where to listen; there were just too many other groups at Groom Lake clouding the conversations. Also, the EPC comms could not be exploited unless they knew when and where a person was using them. In that case, they could listen to the thoughts of the person sending or receiving a message, but the message itself could not be intercepted.

Beyond communication issues, the mission was now beginning to be affected by relativistic effects, more specifically time dilation.

The *Calypso* would continue to accelerate until they reached about 4 percent of the speed of light at which time they would then decelerate for the final portion. The exact amount of time dilation was dependent on how long they traveled at speeds approaching significant fractions of light speed. It also depended on the deceleration profile, which was not precisely known. This was because they did not know of all the planetary bodies available to use in braking once they reached the Cervantes star system. A precise knowledge of every planet in that star system would only be known once they got closer. Roughly speaking, they would experience several years' less time than people on Earth, who would experience one thousand two hundred and fifty years over the course of the voyage.

Over time, as the ship gained more and more velocity, the time slippage would become larger. This would amount to a fraction of a day each year. Since the ship started at zero velocity but was constantly accelerating toward higher velocities, the clock difference, at the moment, between the Earth and the ship was only six minutes. Even this small amount of time had to be accounted for in daily comm cycles with Mission Control.

The last issue that would soon affect the *Calypso* was a crew change. The ship had departed the asteroid belt nearly thirty-six months ago. The current awake crew was due to enter hibernation in another twenty-four months, except Terra and those among the awake crew who had children. They would remain awake until their children were adults. Each awake crew consisted of approximately fifty individuals, some of whom had children. Because some of the crew would continue to remain awake until their children reached eighteen years of age, this created an opportunity for knowledge of voyage events to be passed on to each newly awakened crew firsthand.

For her part, Terra would remain awake the entire time so that The Mother could communicate with her as required. Mission Control knew this now. All the proverbial cards were face up on the table.

CHAPTER 35

Nightmares and Threats

TERRA WOKE UP suddenly. She had a nightmare. A monster had emerged from a dark tunnel and was after her.

Riley came to her daughter's bedside. "Terra, it's just a nightmare. You're OK." Though given her daughter's history and her special abilities, she had doubts about what she had just told her.

Terra sat up, sweat streaming down her face. "Mother, it felt so real, and it's not the first time I've had this exact same dream! It means something. There is something out there, in space, and it is dangerous. It wants me!"

The hair on Riley's neck stood up, and she shuddered. "I'm going to take you to the doctor. Dr. Harrison is always available. I'll call him."

Dr. Haywood Harrison was a big *Star Trek* fan, the original series especially, but he liked the others also. His favorite character in any of the remakes was, of course, Dr. McCoy or "Bones." He listened as Riley explained what she felt was wrong with her daughter. For his part, Dr. Harrison listened; then he said, "I'm going to call Dr. Stuart Carroll. I want his opinion on this."

"Dr. Harrison, you are a medical doctor. How could an astrophysicist help Terra?" Riley asked.

"Mrs. Winstead, I'm just an old country doctor. Even so, give me some space on this," Dr. Harrison said.

Stuart Carroll listened as Terra recounted her dreams, or rather, nightmares. When she finished, Dr. Harrison asked if he could speak with Stuart alone for a moment.

When Riley and Terra left the exam room, Dr. Harrison asked, "A 'dark tunnel,' a 'monster.' Stuart, you know this kid is not like other children. Could she be seeing some threat out there? Now I'm just an old sawbones, but could a 'tunnel' have some 'space meaning,' like a wormhole maybe?"

Stuart started to say something then paused and said, "Well a 'wormhole' is a theoretical concept that has not been proven let alone been identified in space, but yes, if dreams have some symbolic meaning, maybe it could mean that, perhaps."

"I know your astronomers are looking out into space every day for rocks that we may run into and such. Have your folks seen anything weird recently?" Dr. Harrison asked.

"We are looking all the time, but we haven't seen anything like that," Stuart said.

At that moment, Captain Simmons came into the exam room, unnoticed, and said, "Sorry to interrupt. Riley just told me Terra was having some extremely specific nightmares about a monster and a tunnel.

Stuart, I just heard you tell our good country doc that there was nothing unusual out there, but that isn't exactly the truth, is it, Stuart?"

"Well, OK, we have detected some large gravitational attractions directly in our flight path, though it is some distance from us now and is not any immediate danger. It's almost like there are some exceptionally large celestial bodies that we can't see, at least not right now. It could be dark matter. Could it be something else, an anomaly no one has ever encountered, like a wormhole? Yes, but it could be many things we don't know yet. We are watching, trying to get more data on it. We'll figure it out soon enough, I'm confident," Stuart said.

"Riley, Terra, Doc Harrison, and Stuart, could you meet me in my stateroom as soon as possible? Riley, can you get Sam to join us? I'm going to call Peter also. See you all in a few minutes," Captain Simmons said.

After they all arrived in the captain's stateroom, she asked, "Terra, has The Mother told you anything recently about a threat to you or the ship?"

"The last thing she told me was in a dream," Terra answered. "She said there were people who may try to capture me or kill me and that I must listen to my instincts to know when there is some danger."

"Stuart, could others know our flight plan? Let's say some other civilization, like the AOW, knows that we are en route to Cervantes. Are we at risk of them intercepting us?" the captain asked.

"Captain, there are many possible trajectories to our destination. We have supercomputers working out the most optimum course, given what we know about gravitational fields that we would encounter. We don't know all those types of details. We are passing through uncharted territory, so it is the best course according to what we presently know. I suppose if the AOW knows what our state of knowledge is about the route to Cervantes, they may be able to come up with our flight plan, so yes, they could know pretty much where we are and that could help them to intercept us, if that is what they wanted to do."

Captain Simmons spoke. "Suppose they know where we are and where we will be at any point in time. Let's further assume they might even still have some insight gained from somehow listening in to Mission Control. Let's change our course to something less optimal. We need to lose anyone who may want to intercept us. Space is big—we should be able to easily do this. Additionally, I want us to stop pulsing space with our radars. We can just rely on our visual assets. Yes, it is incrementally more hazardous, but we should do this, at least for a while. One of my ancestors was a ship captain in the Second World War. To get across the Atlantic, they zigzagged to avoid Nazi submarines. Let's try some zigzagging, but even more so, let's pick a less-than-optimum course and change to that," she said.

"We'll advise Mission Control via EPC that we intend to do all this. Tell them this is now confidential information and not to be shared with the whole Mission Control team. Also, I want to leave some self-propelled, nuclear-armed, gravitationally attracted space mines. They will lock on to and destroy anything that comes out of the area of the gravitational anomaly."

Peter spoke. "Hopefully, they—whoever they are—will not detect the mines since they are stealthy. On the other hand, we may be starting something between us and the AOW, if that is who 'they' are. Are we OK with that, captain?"

"They, through the actions of Alex, have already tried to kill us all!" the captain shouted.

They all nodded their heads at that.

"We are already in a state of de facto war, and they fired the first shot, so to speak. We are going to do everything we can to protect Terra and the ship!" the captain added emphatically.

CHAPTER 36

Mysteries of Life

TERRA AND HATCHER were eating at the quarters of Hatcher's parents. They each had homework to do, though Terra's were college-level courses. Terra loved learning and reading; Hatcher learned things by doing them. He was now an apprentice technician working with the electronic techs who kept things working on the ship. At thirteen, going on fourteen, both kids were just sprouting their adult wings. Terra was a good listener; Hatcher was full of life and fun. They were a good complement for each other. There was no denying that they liked each other; they studied together every day, and on most days, they spent some time in the observatory sky watching. Often, they would watch old twentieth- and twenty-first-century series. The late twentieth and early twenty-first centuries were the golden age of cable TV. Hatcher loved *The Walking Dead* series and had already watched it three times. Terra was shocked at the gore but slowly tuned it out and enjoyed the storylines more. Terra wondered what she would do when Hatcher and his family went into hibernation. She would miss him immensely. But that was still several years from now.

It was expected that there would be children born during the voyage. That meant that hibernation schedules were subject to change as young children were not put into hibernation. If children were born, the parents would stay awake and raise their children to eighteen years old, the minimum age for safe hibernation. Terra thought about this option, wondering if she and Hatcher would end up together and have children. It was a way to spend more time with him.

The ship had thousands of frozen embryos on board. When they reached their destination, everyone would have children to raise. Larger families were going to be in vogue again. With the normal lifespan of about one hundred forty years, adults would be in the business of raising children for many decades. Everyone on board had been chosen for, among other things, their child-rearing abilities. Studies of colonization on various continents on Earth showed that the success of colonies depended upon having a critical mass of people. Sociologists had long since concluded that children were far from a burden but instead were the key to happiness and success for the new colony they would establish.

Genetic technology had progressed greatly, and the people on board, like most of the people on Earth, had genes that did not have the maladies that plagued life on Earth in the twentieth and twenty-first centuries. Heart disease, diabetes, some cancers, and other undesired genetic handicaps were unheard of now. They had even recently eliminated male baldness! The world of the twenty-fourth century had some particularly good aspects to it though the planet itself was dying. None of the efforts undertaken had been able to resuscitate the web of life that Earth once had. There were too many variables, and even supercomputing codes could not develop successful, balanced, ecosystems *that had appeared* to develop naturally through millions of years on Earth. It was expected that life in a new world would adapt and change as it had on Earth, but no one could predict how life would change as human beings arrived in a new ecosystem. Even with all the great scientific and technological advances leading up to the twenty-fourth century, how life evolved on Earth was still so much a mystery. But it is the nature of all great mysteries and questions that sometimes answers come from the most unlikely sources.

CHAPTER 37

Cat and Mouse

THE SPACE MINES left behind in the vicinity of the gravitational anomaly began pinging, indicating that they had acquired a target and were in terminal pursuit. The pinging suddenly stopped after only a few seconds, far short of what was to be expected if they had successfully locked onto and destroyed a target. The *Calypso* had received the pings because they knew the frequency their transmitters operated on.

Captain Simmons and Gunner sat at the weapons console, silently considering what might have happened.

"We had pings from three mines. They should have pinged for up to an hour while in pursuit of their targets, not just for nine seconds. They were invisible to radars. They couldn't be tracked and destroyed. What happened, Gunner?" Captain Simmons asked.

"My guess is that they were destroyed somehow. Whoever came out of that region of deep gravity is still alive and well. It was a good thing we changed course and skipped town!"

"Yes, but we are punching above our weight class. This means we must do everything we can to be invisible," Captain Simmons responded. "I want to power down the ion drive, go dark again for a few days. I'll let everyone know we are going to be in zero g for a while. Gunner, get Stuart to have his people look back at the area we just transited. I want him to use his 'passive' tech—infrared, visible, whatever he has. I want to know what we can see, if anything."

"Copy that, captain."

Later, the captain and Stuart spoke. "The only thing I can see is from gravitational scans, captain," Stuart said.

"What do you see on the gravitational scans?" the captain asked.

"They showed that there is a massive stealthy object, not a 'natural object,' that transited out of the gravitational anomaly. This is probably some kind of spacecraft," Stuart added.

"What is their course, where are they headed?"

"It seems to be that they are tracking our original course before we changed it. One other thing, Captain. They were traveling at twice our velocity—we would never outrun them."

The captain sighed and turned toward the star viewer on the wall. "We're in trouble. They are faster, and our stealthy mines were ineffective. I doubt we would ever get close enough to use our rail guns. This threat is real, and we need help. I'm going to get the Winsteads up here. I want to talk to Terra."

"Terra, can you talk to The Mother, tell her about our problem, and ask her where that wrinkle goes? I'm wondering if that is our only option—to sneak into the wrinkle and lose the enemy ship," the captain said.

Stuart spoke up. "Captain, we don't know what it will be like to pass through that wrinkle. Perhaps it will destroy the ship. Objects that traverse the wrinkle may need special modifications or unique technology to do it safely."

"We may have to take that risk. The current threat is an existential risk," the captain answered.

Terra spoke up. "I'll ask her. I'd like to go back to my cabin and contact her there. I don't know how long I will be, perhaps less than an hour."

As Terra finished singing, The Mother appeared in her bedroom. "Mother, a ship is chasing us. It came from a wrinkle, we think. Our captain says our ship can't outrun their ship, and we can't destroy it with our weapons. Captain Simmons thinks the only way to save ourselves is to sneak into the wrinkle, but we don't know where it will take us or if it is even safe to go through it. What should we do?"

"Go through the wrinkle. It will put you just two hundred fifty years from my planet. The ship chasing you is our common enemy, the AOW. They want to capture you, if they can, or kill you. They will not expect your ship to turn around and enter the wrinkle. Once you get through the wrinkle, the star field will be different. You can locate your position by searching for known stars. This will allow you to chart a new and much shorter course to our world," she answered.

CHAPTER 38

The New Expanse

THE UNIVERSE EXPANDED after the Big Bang at speeds far exceeding the speed of light. The limitation of traveling no faster than the speed of light only applied to objects within space-time; it did not apply to space itself.

The expansion of the universe was first theorized by Alan Guth and a Soviet physicist in the 1980s. He called it "cosmic inflation." The inflation of the universe could be thought of as a sheet almost instantaneously unfolding itself on a bed, except that there were some wrinkles left in the sheet or space-time as it expanded, stopped suddenly, then rebounded, producing ripples or wrinkles. Those wrinkles, if pierced, could allow a traveler to cover a greater distance than simply traveling across the flat expanse of the sheet. They were natural shortcuts to distant points in space.

The *Calypso* had just traversed one of these wrinkled areas.

The unusually high gravitational fields detected were due to the massive stars located just beyond the wrinkled zone. They could not be seen, but their gravity could be felt on the other side of the wrinkle. This was also an explanation for some of the "dark matter" that permeated space. The effects of dark matter could be felt, but the matter could not be seen. The dark matter was just simply unseen stars on the other side of the wrinkles. Their light was not visible due to the extremely folded nature of space in the area of the wrinkle. The light was redirected by the fold in space and not visible.

The new star field that was visible to the *Calypso* showed that the Cervantes system was now only two hundred fifty years' distant at their current velocity. Captain Simmons had Stuart Carroll calculate a new course to The Mother's planet. A slightly less-than-optimum course was selected in case they were ever followed by the AOW. They hoped that would disguise their location to the AOW to some degree. Captain Simmons also asked Stuart to chart areas of high gravitation that appeared anomalous. These areas were going to be of great interest in future space travel.

Mission Control had been previously told there would be no more positional updates due to the threat from the AOW. They didn't even know the *Calypso* had passed through a wrinkle in space and that their ultimate destination was now much, much closer. The threat from the AOW ship had been evaded. It appeared that nothing was passing through the wrinkle in pursuit of them, and their remaining voyage duration was now only two hundred fifty years.

CHAPTER 39

Giza Necropolis, Egypt, the Earth, 2602 BCE

THE ANTIKYTHERA MECHANISM was discovered in the remains of an ancient Greek ship in 1901, in the waters off the Greek island of Antikythera. The mechanism consisted of sets of planetary gears that could be moved to give the exact predicted position of the sun, moon, and other celestial objects years in advance. It was called the first example of an analog computer. It is believed to have been constructed between 100 and 200 BCE. Nothing like it would be built again until the fourteenth century CE. The dating of the device was off by thousands of years. This was not the only amazing mechanical device of the ancient world, which was far ahead of its time.

Lute measured the new block of stone himself this time. The last stone had the wrong base dimension. This one had the correct measurements. As the chief designer of the Great Pyramid, he had carefully created drawings showing the pyramid in three orthogonal views. It was believed that such drawings were first used by a Roman engineer in fourteen BCE, but they were not fully utilized until the eighteenth century. Again, the experts were off by thousands of years.

Lute had knowledge of these types of drawings from ancient writings dating to the time of The Mother, the first faroh, who had lived eons earlier, in a location far distant from Giza. It was believed that she had vast knowledge that she gained from the ancient Sky Lords. Some of that knowledge had survived the thousands of years since her time. Possession of her writings had given Lute powerful knowledge that surpassed the value of great treasures of gold or silver. Lute had parlayed this knowledge into a position of power. He was the chief designer for Khufu or Cheops, the powerful pharaoh believed to be a descendant of The Mother, the first *faroh* or pharaoh, as they were now known. Lute had other knowledge as well. He knew who the ancient Sky Lords really were.

He had first met Kaanon when he had offered a reward for anyone who could locate deposits of pink granite. A man had found his way to Lute's house, claiming to know the locations of such deposits. The man identified himself as Kaanon, a resident of the Great Rift Valley in Ethiopia, though he was not originally from there. He had a strange accent and language. Lute overheard him talking in this strange tongue on one occasion in his room. Lute had hit it off with him in their first meeting. Kaanon was smart and shrewd, yet easy to have a good relationship with. He also had a sharp sense of humor.

Lute offered to let the outsider stay at his house while they assembled men and material for their expedition to the sites of pink granite that Kaanon identified. Lute and Kaanon would talk about many subjects late into the night as they drank wine. Kaanon was full of knowledge and stories. One night, after Kaanon and Lute had become well lubricated with wine, Kaanon claimed he was originally from across the great sea, in a place he called Eden. Kaanon also said that he had met and knew the Sky Lords.

Lute was skeptical and asked, "Who are they? What land are they from?"

"They are not from our world. Come outside, I will show you where they are from," Kaanon answered.

As they walked into Lute's courtyard, Kaanon pointed to a star in the northern sky. "There, see the faint one?"

"What do you mean they are from 'there'?"

"They live on a world like ours that circles that star."

"How did they get here? Surely it must be a great distance," Lute asked incredulously.

"They had a great starship that sailed the night sky for many moons to reach our world."

"Kaanon, I find you to be a man of humorous intentions! No one can reach the heavens, let alone move around among the stars!" Lute answered, slurring his words and nearly falling over in laughter.

"Even so, Master Lute, they, in fact, did just that. Let me show you some of their artifacts."

From his bag in his room, Kaanon brought back an item that looked like a metal cylinder. He pushed a button, and a light came out of the end of it.

Lute smiled. "What is it called?"

"A flashlight. It runs on electricity, like the lightning in the sky."

Lute's laughter slowly tapered off as he became serious. "You have other artifacts in your possession?"

"Yes, I have just a few with me. I have others hidden away in a cave in the Great Rift Valley. I will give you one as a gift. It is a device that can predict when the sun will be darkened in the day, as I am sure you have seen in your days. The pharaoh will think you are a wise man when you make such predictions many moons before it happens!"

"Kaanon, you knew the first pharaoh, 'The Mother'? Did she not live long ago?"

"Yes, she lived a great many moons ago, long before people lived in this land."

"How is it that you could know her? You look to be even younger than me!"

"I am older than any of the pharaohs who have ruled in this land. I do not grow old."

"I would say this is not so, but your artifact causes me to think you may speak the truth. Why do you not grow old?" Lute asked.

"I displeased The Mother. She sent me away from my people and changed me so I would not age," Kaanon answered. "This was my punishment—to live in this world and suffer until she sees fit to bring me to hers. I am to tell her of the doings of men in this world through another device I possess. It lets me talk to her, even on her world, far up in the night sky."

"It seems to me that to live such a long time is no punishment but a blessing from the Sky Lords," Lute said.

"Trust me, Master Lute, to live in this world and experience all the suffering we see is no reward at all. I long to be with The Mother in her world. Someday I will be there."

Lute kept several of Kaanon's artifacts after Kaanon strangely disappeared many years after they first met. One of these artifacts was the device that divined the position of the planets and other celestial bodies. The other one of note was the device that was nearly identical to the device Kaanon used to talk to The Mother. Kaanon had not said who he had spoken to on that device. Lute found it in Kaanon's bag in the days following his disappearance. These artifacts were kept in Lute's family for millennia. When Ptolemy first built the Great Library in Alexandria, Lute's descendants sold the devices to the Greeks so that they could be safeguarded in their library, along with many of the papers of The Mother, which also ended up in Lute's possession. Legends of The Mother became so prevalent in Egyptian society that she attained the status of a goddess named Isis, the most popular of all Egyptian deities. Lute was taught The Mother's language by Kaanon, and he taught it to his firstborn son. But with the passing generations, knowledge of The Mother's language was lost although the knowledge written by her in the papers remained, having been carefully copied many times through the centuries.

The Great Library of Alexandria was destroyed around the year zero. Legends of what happened circulated for centuries. None of those who witnessed the destruction of the library would ever forget what they saw though their accounts were dismissed in the ensuing centuries. The accounts told of "fire from heaven" and a "great mushroom cloud." Nothing of the great works stored in the Great Library were ever found. In fact, no remains of the library's structure were ever located, so thorough was the destruction.

It turned out that not all the devices sold to the library were destroyed. Weeks before the destruction, a Greek ship owned by a wealthy Macedonian merchant left Alexandria with two items. The ship was caught up in a great storm and sank near a Greek island. The device that predicted the position of heavenly bodies sank with the ship and was found again in 1901. The other device, the communicator, was carried by survivors of the shipwreck to the island of Crete and was lost until the twenty-third century CE. This was the device Ted Bywater used to contact the AOW.

CHAPTER 40

The Moon, 2582 BCE

KAANON WAS LED into a small room. He was still very uneasy with the small amount of gravity, so he walked very gingerly as he entered the room. He had been picked up yesterday at dawn. He had been told that this would be how The Mother would pick him up when she determined he was "ready." The reality of that finally happening, after the unbelievably long time he waited, was still roiling his reality.

"Kaanon, be seated," said a man across the table from Kaanon. "I represent The Mother, who now lives on the world of the Sky Lords. The Mother enhanced your genetic reproducibility and made you age very slowly so that you would learn from your actions. You are being offered the opportunity to stay and assist in the development of this world, or you may return to the world of the Sky Lords and be with The Mother of this world."

"What manner of work would you have me do if I stayed here?" Kaanon asked.

"The people of this world are entering a new phase of their existence. They are beginning to form into large societies, even empires. You know this—you were in the Land of the Pharaohs. For them to learn from successive empires, they need to have written records. Commonalities among groups are also needed. The large societies need shared languages, history, mythic beliefs, and rules of personal conduct. The last items, mythic beliefs and rules of conduct, are critical. This would be the nature of your assignment—to help in the creation and proliferation of mystical religious beliefs. Are you interested?" the man asked.

"Yes."

"Good. You will be given the means of traveling between here and the Earth. You are not to reveal your identity and purposes to these people. You will report to The Mother on the happenings among the people and will be given directions on what to do to fulfill your mission. There are others here and on Earth who will work with you on these tasks. You have done well and learned from your actions. You can be of great service to The Ones," the man said.

CHAPTER 41

The Original Benedict Arnold

KAANON RETURNED TO the Earth later that year in a panic. He was not sure what had happened to the EPC device the stranger had given him. He had not dared take it with him to the moon when he was picked up. When he had gotten back, it was not in the cave in the Great Rift Valley where he had hidden other items of value. Now, he could not tell the stranger, who had become his friend and business partner, of his new assignment for The Mother. This would surely be of interest to his people. He now had enormous influence concerning the happenings on Earth, and he was an important part of The Mother's team.

He had met the stranger, Ish, two years earlier, as he traveled back to his home in Ethiopia. It was night, and he camped alone on a mesa about two days from the Rift Valley. He was sitting beside his campfire when a stranger walked right into his camp and sat down opposite him.

"Kaanon, we finally meet! May I join you? I have fine wine to share," Ish said.

"I am sorry, do I know you?" Kaanon asked, wary and a bit frightened of the forwardness of the man.

"No, we have not met, but I have been watching and following you for a long time. I feel as if I know you. We both are loyal lieutenants, valuable to our superiors, but vastly underappreciated. I am Ish, a son of Elotim, someone we both know. I, like you, have made mistakes, and for that I was marooned to this backward little rock for quite a while. But I reformed myself and was accepted back into the fold, as you will be someday—soon, in fact, according to my sources," he said.

Kaanon looked directly at him, thinking, then said, "Wait, I know of you. You were the one who tried to kill The Mother and was banished, am I correct?" Kaanon paused then continued before Ish could answer. "What do you want?" He was still frightened, but now he was also curious.

Ish drew in a deep breath then continued, with slow deliberation. "Yes, I am that man, to answer your question. I want to have what I deserve, Kaanon. I want to run this world and eventually other worlds as well. I am building my team, and I need a few good men, like you, to do it. There is great power and wealth to be shared in this. Are you interested?"

Kaanon stared at him, saying nothing for a moment, then answered, "Yes, yes, I am. How can I be of assistance?"

"I will let you know when the time is right. For now, I want to give you a device on which we can communicate.

Protect this—do not lose it. If you do, you could endanger my plans. Here, take it. You know how to speak and write in the old tongue, right?" Ish asked.

"Yes, I do."

"Just type in the letters of the words you want to say and press the green button to send it to me," Ish said.

"Can do," Kaanon answered.

CHAPTER 42

Plotters

KAANON WAS BESIDE himself with frustration. "Shit! Where is the damn thing?" He had searched the cave for the second time. *Ish will kill me for sure*, he thought. Now he had no way to contact Ish. He could not even tell him he had lost the device. All he could do was wait and see if Ish showed up in person to ask him where the device was, presumably after he failed to respond.

It turned out he did not have long to wait.

Several days into his return journey back to Giza, Ish showed up again at Kannon's evening campfire, as he had done the first time they met. Their conversation did not go well.

"So right after I say, 'do not lose this important item,' you lose it. You have no idea where it is or who has it!" Ish shouted.

"No, Ish, I do not know where it is, but I am still looking."

"Never mind the device, I will give you another. Needless to say, don't lose it," Ish said.

"Of course, I will guard it with my life," Kaanon responded.

"You will guard it with your life, or you will forfeit your life." Ish looked at him and smiled wryly. "Now let us talk about who the people are that The Mother has on Earth and on the moon. Are there any that can be turned against her? Any with grudges or unmet ambitions?" Ish asked.

Kaanon thought for a long minute. "Yes, a man named Kush, who is in Canaan. He and I have met only once, but we really hit it off. Both of us feel that we have more to give than we are giving in our current capacities. The worship of Isis, a Mother-like deity, is extraordinarily strong in the Land of the Pharaohs. The Canaanites are a people prone to copying others, especially regarding mystic beliefs. They will adopt the worship of Isis. If we can prevent this, we can prevent the ascendancy of females as supreme deities in Canaan and Mesopotamia. This will dilute the power of The Mother for millennia and promote the development of large aggressive, male-dominated, autocratic regimes, ones that will make it easier for us to control large groups of people."

"Contact Kush, get him on board with this plan. You do this, I will forgive you for the loss of the device," Ish shouted.

CHAPTER 43

Strange Lights in the Heavens

PETER STOOD LOOKING at the screenshot of the object Stuart showed him. "Stuart, what am I looking at?"

"It is a red dwarf star," Stuart answered.

"It's directly in our path. I assume we should deviate around it?"

"Yes, we are reading a decent gravitational field out to a couple of AUs or more, so yes, we do not want to wrestle with the alligator, small though he may be." Stuart laughed. "Though, I must say, I do not know what that would be like. I've never seen one. They went extinct before I was born!"

Peter looked at him and shook his head. "You physicists have such an odd sense of humor. I will let the captain know we will be making a course change to deviate around this thing."

"Oh, Peter, one more thing. There is an Earth-sized body orbiting this thing. It has lights with colors. Too bad we could not send the shuttle to check it out. It is rotating around the red dwarf with our direction of velocity. I think a shuttle could race ahead, land, get back off the rock, and speed up to meet us for a rendezvous. You know, our mission is exploration also," Stuart said halfheartedly.

Peter said nothing for a moment, then answered with a simple, "Hmm, let's ask the captain."

Captain Simmons and Gunner listened as Stuart and Peter explained the plan.

"We take a shuttle, accelerate using the shuttle's ion engine for five days toward the planet—I think we are calling it a planet. Anyway, that will allow us to land there and spend up to twelve hours on the surface. When we are finished, we can leave. With the velocity of the planet, we can accelerate to match *Calypso*'s velocity within two days. But we must initiate this plan within forty-eight hours for it to work," Peter said as Stuart nodded his head in agreement.

"Everybody, hold on," Captain Simmons paused and continued. "Why not send a probe? Why risk our people?"

"Because there is more," Stuart said. "Something Peter didn't mention."

"What's that, Stuart?" the captain asked.

"There are very powerful lights on the surface, and they are blinking out a message." Stuart added, "We think it is some version of the old Morse code."

"Wait, what? What does it say, the message?" the captain asked incredulously.

"Well, that is where we can't say what. It is a series of short and long blinks of the lights, but it does not correlate to our languages on Earth," Peter answered this time.

"It might be an SOS," Stuart chimed in.

Peter shifted on his feet nervously then answered, "We don't know that, Stu."

The captain sat down in a chair and leaned back, putting her hands on her temples, rubbing them. "I have a headache. This whole trip has not been what we expected. Mothers, messages, saboteurs, and lights! What else is in store for us?" she asked with half a smile. "Our mission is exploration, you know, to 'boldly go where no man has gone before'! OK, we will go. Next question, who goes?"

"Me," Gunner said. "I can pilot the shuttle."

"Me too," Stuart said. "I'm a planetary specialist also, in addition to an astronomer."

"We probably need a third person," the captain said.

"Stephanie," Gunner said. "She should go also. She's qualified on shuttles."

"OK, make it so," the captain said, smiling slightly again at another *Star Trek* reference.

* * *

The captain, Peter, some assorted crew members, and the shuttle crew assembled in the shuttle bay prior to leaving.

Peter briefed everyone one last time. "OK, you all have two weeks of rations. Twice what you should need. I have included electrostatic guns, just in case, and you have two old-fashioned Musk space-certified firearms. You folks should have everything you need."

"We'll be fine, 'Dad,'" Gunner said, laughing heartily.

"If you don't feel right about anything, you abort," Captain Simmons sternly said.

"Will do, captain," Stuart answered.

As the three walked to the shuttle, the assembled crew gave them a rousing cheer. All three waved and walked up the ramp to the shuttle entrance.

The crew emptied out of the shuttle bay and watched on the viewer screen as the huge shuttle bay doors slid open, revealing the black expanse of space, interspersed with stars. The shuttle thrusters ignited, lifting the giant craft off the floor, and then pushed the shuttle out of the *Calypso*. Within a minute, the main chemical thrusters ignited, and the shuttle sped out of view. A few minutes later, the crew watched via a camera feed from shuttle cameras as the ion drive ignited. The shuttle vibrated slightly from the thrust. They were on their way.

CHAPTER 44

Landing

PRIOR TO ARRIVAL at the planet, the AI planetary software selected an appropriate landing site, which the three agreed looked safe but also interesting.

"*Armstrong*, how do you read?" Sam asked.

"I read you five by five, *Calypso*," Stephanie answered. They were using light speed comms that came with a slight delay that made communication a bit halting and slow. "We are initiating a deorbit burn. Burn initiated, beginning descent."

Inside the shuttle, the three looked at one another as the shuttle began shaking violently. "*Calypso*, we are experiencing a significant atmosphere. We will be initiating the spinner. We can save fuel," Gunner said.

"Copy that, *Armstrong*," Sam replied.

The spinner acted like helicopter blades, providing thrust in an atmosphere environment. For anyone watching the shuttle, it would have looked like a giant spinner toy, popular in the early twenty-first century. It had a center section, three blades connected to an inner ring or raceway, and an outer ring. The blades were connected to an inner ring raceway that allowed them to spin while the center section or the cockpit and crew section did not rotate.

The shuttle vibration dropped significantly as the craft used the atmosphere like a helicopter to control its descent.

"Descending through eight thousand meters altitude. The radar shows rolling hills, approximately sixty meters in height. Visual is partly obscured by gaseous clouds. The light is dim. The surface might be brownish or gray. I see a white area, a frozen liquid, maybe water. Whoa, I see shapes, could be structures. They might not be natural features! I am also reading elevated radiation levels, not lethal, that would be equivalent to one thousand millirems exposure for a normal adult male," Stuart said, out of breath and obviously excited.

"Landing legs deployed. Landing lights on, monitoring descent camera. No obstacles, I'm going to take us straight down. Forty meters, thirty meters, twenty meters, ten meters, nine, eight, seven, six, five, four, three, two, one . . . contact! *Calypso*, touchdown confirmed," Gunner relayed through his mic.

"I am initiating the planetary sampling system," Stuart said.

All three crew members sat silently while the automated system sampled the air and soil. Within ten minutes, the preliminary results were in.

"*Armstrong*, this is *Calypso*. We are reviewing the samples from the planetary sampling system. We see a mostly nitrogen atmosphere. Oxygen 15 percent, other trace gases and water vapor are at extremely low levels. Temperature negative thirty-five Celsius, no wind, soil mostly oxygen, silicon, and iron. Background radiation unchanged from levels noted during descent. Wear environmental spacesuits during EVA (extra vehicular activity) over."

"Copy that, *Calypso*," Stuart said.

"OK, the weather could have been worse. It's like being on a mountaintop in winter. We want to stay out no more than six hours," Stuart said to the other two crew members and to the *Calypso*.

"Copy that, *Armstrong*, we concur," Sam said.

"*Calypso*, what are the decontamination rules here?" Stuart asked.

"Reenter through quarantine compartment, incinerate environmental suits. Seal off equipment that is used on EVAs in the cargo bay. For yourselves, perform standard watch and observe. Do two blood draws with bioscreens twenty-four hours apart on your return trip," Dr. Harrison answered.

"*Calypso*, the light sources we saw are about four kilometers west of our position. We are going to head there first. Gunner and I will do this first EVA. We will use the crawler for transport. We will stop short and perform observation with the aerial scout drone, or ASD, then proceed closer if we feel good about things," Stuart said.

"Steph, I know you want to go out, but we need you here in case we need support from the *Armstrong*. Use the second ASD to find a closer landing site while we are in transit in case we need you to move the shuttle for a quicker pickup and exit," Gunner said.

"Copy that," Stephanie answered, as she sat back down.

"Gunner, launch the ASD. I'll monitor it as you drive," Stuart said.

Once they got outside, the scout drone was ejected from the front of the crawler and immediately began an autonomous recon of the route they were going to take. There was a hill between them and the lights that they would need to traverse.

"The ground seems firm, with just a light covering of dust. I've got the ground-penetrating radar going. It shows firm soils in front and all around us. We should be OK to proceed at max crawler speed," Stuart said.

"Copy that," Gunner responded.

Within twenty minutes, they were two hundred meters from where the light sources were located, except there was nothing to see—no lights, no structures, just flat ground with no boulders, nothing.

"Well, jeez, Stu, they didn't roll out the welcome mat and red carpet for us, did they?" Gunner wondered.

"Let's get a little closer," Stuart said. "Then get out and walk on foot. Maybe everything is underground?"

"Oh, hold on, the radar is showing something sixteen meters below us, a cavern of some kind, but where's the entrance?" Stuart asked, bewildered.

Just then, the crawler tilted to one side and started to sink into the ground.

"Stuart, get out now!" Gunner shouted as he jumped out and ran away from the crawler.

Stuart was not as fast; he jumped out but tripped and fell a few feet away from the crawler, feeling a snap in his left foot. "God damn it, that hurts!" Stuart said as he got on his hands and knees and tried to get to Gunner.

The sand was moving and sinking, taking the crawler and Stuart with it.

Gunner shouted into his comm, "Steph, peg our location! We are sinking into the sand, maybe an entrance to a cavern. Get to our location, stat!"

Gunner ran to Stuart. He knelt and tried to get him up and away from the sinking ground, but he was sinking with Stuart and the crawler. The ground gave way, and they both dropped quite a distance, coming to a stop when they landed on their chests, their heads bouncing up. Both were unconscious.

CHAPTER 45

Injured, Lost, but Not Alone

STUART WOKE UP first, turning his head to one side to look at what was around him.

Low pinkish-white light illuminated the walls. Stuart turned his head to look for Gunner. He saw him face down, not moving.

"Gunner, do you copy? Are you OK?" he asked.

Just then, Stuart noticed that his face shield was shattered, and his comm headset was lying on the ground next to him. He realized he was no longer breathing his suit's oxygen. Panicked, he tried to stand up but collapsed from pain in his chest and legs. He remembered his injured left foot, but he did not remember his chest, legs, and head being injured also. He realized he probably had broken ribs and maybe a broken leg. Possibly a concussion as well.

"Damn, my head hurts!" he said aloud, but the sound seemed to die as it hit the pinkish-white walls. As he spoke, he spit blood out of his mouth onto the floor. The floor seemed warm, slightly soft, spongy. *Maybe the spongy floor saved me from death? I fell quite a distance*, he thought.

Just then, Gunner moved his arm. Stuart reached out and grabbed his gloved hand and squeezed it to let him know he was there.

Gunner squeezed back, exhaled, then said, "Gawd, I feel shitty!"

"Misery loves company. I feel crappy also," Stuart responded.

"Where are we?"

"Underground, perhaps in the cavern I saw on the radar."

"How long have we been here?"

"I think about thirty minutes. We are both not breathing suit air. I think I have broken ribs, maybe a broken leg, plus my head hurts like hell!" Stuart answered.

Gunner responded, "I believe I have a concussion. But I think I can stand up. I am going to try."

"Be careful, take it slow," Stuart cautioned.

Gunner raised himself onto his hands and knees then brought one leg up, then the other and stood upright. "I'm a bit dizzy, and—" Gunner stopped suddenly then turned, retching, and threw up. "I think I feel a bit better now."

"Maybe you do, but that smells awful!" Stuart smiled weakly as he answered.

"Let me have a look at you, Stu."

Gunner knelt and began examining Stuart.

He felt along his lower calf and said, "Yeah, you have a ruptured Achilles tendon. I can feel it all rolled up behind your foot. Your legs do not look obviously broken.

I mean, you don't have a compound fracture, thank God! Let me see if I can roll you over, get a look at your chest."

As he rolled him over, Stuart cried out in pain. "Well, you might have a broken rib or two. Your rib cage has some bumps that it shouldn't have."

Gunner stood back up and walked slowly and painfully over to the illuminated wall. He reached out to touch it, and the wall slid open. Gunner jumped back, wincing in pain as he did. A person stepped forward and said something in a language he didn't understand. The person, a woman, repeated the same unknown phrase. Gunner stared at her, and she stared at him.

"Ma'am, we are both injured. Can you help us?" Gunner pointed to his head and then at Stuart's foot and chest.

The woman said something softly then walked over to Stuart and held his foot. Stuart cried out in pain. The woman put her hand over Stuart's eyes, and he fell unconscious. She then put both of her hands around Stuart's injured foot and gently massaged his calf and leg for several minutes. She moved her hands over his chest, pressing deeply as she did. Lastly, she held his head in her lap as she sat by him and moved her hands over his head and hair. Finally, she placed her hands over his eyes again for several seconds and then removed them.

Stuart opened his eyes and said, "Gunner, what happened? I must have passed out again, but I feel much better. So strange."

The woman smiled. Stuart turned his head and saw her holding his head and jerked away.

"What did she do to me?" Stuart blurted.

"Dude, I think she healed you. She fixed you up somehow," Gunner answered.

Stuart rose to his feet and faced the woman. He was silent for a moment as he stared at her. Then he said, "Thank you."

She smiled and then went over to Gunner and pushed gently on his shoulders while motioning for him to sit on the ground. Gunner sat. The woman sat down behind him and gently pulled on him, indicating he should lie down flat on his back. After he did, she put her hand over his eyes, and Gunner went unconscious. She rubbed his head and scalp with both hands as she had done to Stuart. When she finished, she put her hand over his eyes again. He opened his eyes and awoke.

"Wow, I feel better. I feel great, as a matter of fact," Gunner said.

The woman smiled again. She motioned for them to follow her down a hallway behind the wall that had slid open.

CHAPTER 46

Shuttle *Armstrong*

"CRAWLER ONE, COME in. Gunner, Stu, please respond!" Stephanie said. All she heard in response was static. "*Calypso*, did you hear any of what was said by Gunner?"

"*Armstrong*, yes, we heard. Any more information?" Sam asked, back in ComCent with the captain and Peter listening behind him.

"I'm going to take the shuttle to within fifty meters of their pegged location," Stephanie said. She hoped that Gunner and Stu might have heard this also.

Stephanie retracted the cargo ramp and lifted off the surface using the spinner blades. She followed the westerly course the crawler had traversed. When she found a suitably flat location, she pulsed the ground repeatedly with the powerful landing radar to check for ground stability. Satisfied that the ground would support the weight of the shuttle, she gently landed.

She went down the stairs to the passenger and cargo area to prepare the scout crawler drone (SCD) for an EVA to the location from which Gunner and Stu had last called. She went back up the stairs to the cockpit and brought up the SCD cameras. They seemed to be working, so she remotely drove the crawler to the cargo air lock. Once outside, she directed the crawler toward the pegged location. As the SCD neared the last known location, she pulsed the ground with its ground-penetrating radar.

"*Calypso*, *Armstrong*. I see what seems to be an open cavity about fifteen meters or so below the surface. No sign of Gunner, Stu, or their crawler. There is a shallow depression at the pegged location, with what seems to be some ground discontinuities. I am guessing the location is an entrance to the cavern. It opened, and that is where they are. The crawler tracks lead to the pegged location, but they stop abruptly. Whatever happened to them happened right there, over."

"Has the presence of the SCD triggered an entrance to be opened up?" Peter asked.

"Negative, the SCD is currently stopped, and no activity can be seen," she answered. "I am going to try comms again and see if they respond. Gunner, Stu, do you copy? Status? Are either of you injured?" There was no response.

"*Calypso*, any ideas? I could put the drill set on the SCD, bore down to the depth of the cavern, and get a camera down there. What do you think? Over."

"*Armstrong*, this is Simmons. You are almost three hours on the surface now. We must keep pressing ahead to get you all off within twelve hours. Proceed with the drilling."

"Stephanie, this is Peter. Use a twelve-centimeter bore drill. You should be able to get a minicamera down a hole of that diameter, over."

"Copy that, Peter," Stephanie answered.

The drill began operating. The ground was extremely hard, and it made slow progress. Thirty minutes later, the equipment had only drilled down a meter.

"*Calypso*, *Armstrong*, this is going too slowly. The ground is extremely hard. I am thinking we could try an explosive in the hole, see if that loosens things up a bit. Thoughts? Over."

"Stephanie, I concur. You should have bore hole charges on board. Let us try the one-kilogram charge. If that doesn't work, we could try a larger explosive charge. Over," Peter said.

"Copy that, Peter. I am going to have to go out to the drill set to put the charge in the hole. I am going to tether myself with a cable to make sure I do not fall into the cavern unexpectedly, should the entrance open again. This could burn another two hours for me to do this. We would be at about six hours by the time the charge goes off. I may want to move the shuttle farther away to keep it safe from anything that might open. Also, blowing a charge may piss off any ETs who might be in that cavern. I want some distance to give me time to lift off if I need to."

Stephanie lowered the charge into the borehole. She attached a transmitter to the wires coming out of the hole and moved the SCD back fifty meters. She got back into the shuttle and lifted off, moving it back to the original landing site. As she prepared to detonate the charge, she looked at the SCD camera views to see if anything had changed. Something caught her eye, just as she was ready to detonate the charge, causing her to stop.

The ground in the area where Gunner and Stuart's crawler had been parked seemed to shake, sink, and then move, or rather slide open. She watched as debris fell into a deep hole as the ground slid northward, exposing a dark opening. A few seconds later, the hole seemed to lighten up. The ground started shaking slightly again. About a minute later, a flat open elevator rose even with the ground level. Gunner, Stuart, and several other people, dressed differently from Gunner and Stuart, stood on the elevator. Gunner and Stuart waved at the cameras on the parked scout crawler then started walking to it.

When they reached the crawler, Gunner picked up the comm mike and said, "Can you come and get us, Steph? It's freezing out here!"

CHAPTER 47

Mending Broken Bodies and Hearts

WHEN THEY GOT back to the shuttle, Stephanie asked, "OK, Gunner, Stu, what happened out there? Who are these people?"

"The ground opened up and swallowed us," Gunner said with a smile.

"Right, you mean the door thing slid open, and you fell down into that cavern, correct?"

"Yes, I guess the elevator was supposed to have lifted to prevent us from falling, but it didn't work as designed. This place has been here for a very, very long time. I suppose the scheduled maintenance has been lacking. Anyway, we both got hurt badly, Stu worse than me. He had broken ribs and a ruptured Achilles tendon, and we both had concussions. But we are better now," Gunner explained.

"You are better now? How is that even possible?" Stephanie demanded.

"Because of her, their healer." Stu pointed at one of the women in the group.

"Who is she?" Stephanie asked.

"Her name is Simi. She is a master energy healer," Stuart said.

"Right, gotcha," Stephanie said sarcastically. "Can she read my energy also? What about my aura?"

"I know how it sounds. I do, Steph, but she really did fix us up," Gunner interjected.

"Okey dokey. My neck and back are sore. Can she work on me?" Stephanie asked mockingly as she turned and looked at Simi.

Simi smiled and walked toward Stephanie. She attempted to touch her neck, but Stephanie recoiled away from her.

"Steph, let her try. It won't hurt anything or anybody," Stuart said.

Stephanie stopped moving. "OK, my neck is stiff when I turn my head to the right. Sometimes my right arm goes numb and tingles, plus my lower back just aches when I sit for a while." She said this as she motioned with her hand to the right side of her neck and then to her lower back.

Simi moved behind Stephanie and began massaging her neck, right shoulder, and arm. Stephanie closed her eyes and let out a small sigh of relief. Simi moved to her lower back and began pressing in several spots. When she did this, Stephanie jumped slightly then relaxed. Simi finished and came around to the front of Stephanie and asked, "How feel?"

Stephanie looked at her for a moment and then said, "Better. Yes, I feel better."

Simi then said something in her language to a man in the group. The man stepped forward and said in perfect English, "She said she needs to work with you on the pain in your heart."

"What? What do you mean?" Stephanie asked.

"You have pain over a lost loved one. This is causing your body to manifest pain in your neck and back," the man answered.

Stephanie stepped back. "I don't know if I want her messing with my mind!"

Simi stepped toward Stephanie and said in broken English, "It OK. I help you, later maybe?"

Stephanie stared at the floor for a moment and said, shakily through tears, "How do you know?"

"I know. I know, your heart broken. I heal it," Simi answered softly.

Stuart went up to the cockpit and radioed the *Calypso*. "*Calypso*, this is Stuart. We have eleven individuals who are asking for asylum."

"Asylum? What do you mean?" Captain Simmons asked.

"There are people on this planet who escaped from the AOW. They originated on a planet quite distant from the one on which we landed. Apparently, this planetary body was an outpost of The Mothers. An 'emergency harbor,' a gas 'n grocery stop, to use a twentieth-century term," Stuart added.

"Anywho, they came here just before their world was 'acquired' by the AOW. They also have something of significant value. Well, the value is not calculable. We should talk only when we are back on the *Calypso*. That is why I am using the tight beam with quantum encryption for this communication. I recommend we take them on board. They are down to their last food items. They would perish if we do not take them on board. Over."

"Copy that, Stu. Bring them with you," Captain Simmons said. "Remember, Doc Harrison wants all of you to undergo the bio screen blood panels twice before you arrive. Given that you are bringing back 'aliens,' we'll need to keep them and you three in isolation for seven days upon your return. Over."

"Copy that, *Calypso*. See you in a couple of days. Stuart out."

CHAPTER 48

The Map of the Universe

CAPTAIN SIMMONS, PETER, Sam, Terra, and Riley sat outside the window from Stuart, Gunner, and Stephanie. The isolation rooms in the ship hospital allowed for this type of meeting between loved ones and those who were in the isolation rooms. With the curtains drawn back on all three rooms, each patient could see and talk to those assembled to visit them.

"So, Stuart, tell us what happened on the planet, especially focus on this item you referred to," Captain Simmons said.

Stuart began. "Well, they took us to a room, a holding room, I guess. We were in there for only a few minutes when they got me and took me to another room. This one seemed like a conference room. The eleven individuals we took on board were all there, standing around two holographic images. Now, mind you, I only saw this for a few minutes. One image looked like it was the universe—large, relatively flat, with clumps of galaxies. The other image was a zoom in on some part of the universe. I think it was our galaxy, our part of the galaxy.

"The tall younger guy among the refugees, Timeye, faced me and asked, in perfect English, 'Are you from Earth?' I must have stood there for a moment without answering because he asked me again. I just did not expect him to speak English. I nodded and said yes. I noticed on the zoomed-in image that there were irregular wavy lines between stars in some locations. He then asked me if they could come with us to The Mother's world. I said, 'You know, it is still a long journey.' I wanted them to know that. Timeye said they could make it much shorter still for all of us. I asked him if it was because of the wrinkles. He looked at me for a minute, smiled, and said yes. I told him I thought it would be acceptable for them to come along but, my boss would make the final decision. He asked me if I wanted to ask my boss. Anyway, that is why we ended up back on the surface."

"Stuart, tell us what they told you about these wrinkles and more about this device," Captain Simmons said.

"You all recall the wrinkle in space-time that we traversed to escape the AOW?" Stuart asked. There were nods from those outside the rooms.

"The Mother knows the location of all these wrinkles, but more importantly even, how, and where these wrinkles will spread. Space-time spread out *almost instantaneously* very shortly after the IPT—"

"Stuart, language, speak in our language." Gunner smiled as he interrupted.

"Initial place and time, though I assumed—"

Gunner again interrupted, "Stuart, you know what they say about what happens when you 'assume'? It puts an A-S-S in front of U and ME!"

"Everyone, let's stay on the subject!" Captain Simmons said, exasperated.

"OK, um, so the universe, to use the common term, expanded much faster than light speed just after the Big Bang, using the older, more familiar term. This rapid expansion ceased as soon as it began. Now the whole universe is still expanding. In fact, the expansion is accelerating. The initial ripples roll across planets, stars, and galaxies and get reflected by some dense objects, like collapsed stars. The wrinkles, like the one we used, are not permanent features in space-time. These rebounding ripples move the wrinkles and create new ones at various locations and times. Somehow, The Mother had the initial conditions of the Big Bang and the progression of states, locations, and times of the ripples and wrinkles. She can travel around the universe using this knowledge. It still takes time, but for some locations, it makes travel practical in terms of voyage durations. In short, she has a map to get around the universe. The Caldonians, our refugees, preserved and protected a copy of this map for The Mother. Though it is not a paper map, it is of tremendous value. The AOW wants it, and they want Terra and other children like her."

"Maybe it is just me, but you could have just said, 'She has a map to get around the universe,'" Gunner said with a smile.

"What does this map look like?" Captain Simmons asked.

Stuart began. "I only briefly saw it. It looks like a spherical glass ball rigidly mounted on a pedestal. I think it might be a computer with holographic projection capability. It can project a map of the universe, complete with wrinkles and ripples, at any time desired."

"Like Google Earth, only this is Google Universe. I've been meaning to locate where my old girlfriend lives, and maybe this device—"

"Please, for the love of God, Gunner!" Captain Simmons again interrupted, seemingly serious, but with a trace of a smile at Gunner as she looked at him. There were a few muffled laughs; then everyone settled down.

"So they have this thing among the items they brought on board? Do you think they will show it to us or not?" Peter asked.

"I think they might. It calculates courses for them and other such useful things. Maybe it was part of the navigation system on their ship, which, come to think of it, they never told me what happened to their ship . . .," Stuart said, trailing off as he did.

CHAPTER 49

Refugees

"THE NAME OF our world is Caldonia, which does not yield a correct pronunciation, but I have listened to many of your broadcasts, and I believe that is an easy enough pronunciation for you," Timeye said. "From your world to ours, it is approximately seventeen thousand five hundred light-years, according to your reckoning of years. We have been monitoring your broadcasts at remote, unmanned locations, much closer to your world. We then forward these broadcasts via what you call EPC to our world. That is how I know your dominant language. I also can speak Spanish and German.

"We have been using EPC for thousands of years. Undoubtedly, other civilizations use the same method of communication for the safety of their worlds. This is why you have not detected electromagnetic emissions in any of your SETI activities. But we have been watching and listening to your people for an exceptionally long time. We know of the activities of the Sky Lords in developing the singular intelligent species, *Homo sapiens*, on your world. We also know your world is dying. You have almost always chosen male leadership, which has harmed your development. We do not believe you will survive this tragic mistake."

"Do you know of our mission to 'our' Mother's worlds?" Captain Simmons asked.

"We learned of your mission to save a portion of your people through your voyage to a new world. This would have failed—you are not even close to being able to create and modify a world for habitation for the singular intelligent species, what you call *Homo sapiens*, the species of The Mothers," Timeye said. "For your sakes, your Mother has saved your people by inviting you to a suitable world near hers. Currently, you do not have the ability to properly develop a world for your habitation. You should know that this is still beyond our capabilities. The Mothers are the only ones who possess this ability."

"Why did you leave your world?" Captain Simmons asked. "Where were you going?"

"Our world was seduced by the AOW. They appealed to the greed of some of our people. This led to a rejection of our Mother. Greed and self-interest were reintroduced to our world. Wars soon followed the introduction of AOW-styled male leadership. Our planet was also destroyed. We were traveling to the world of The Mothers when the AOW damaged our ship. We were forced to land on the waystation planet where you found us. We would have perished had you not come when you did. We thank you for your assistance," Timeye said.

"So intelligent life developed on your world, like ours, because of the efforts of The Mothers," Captain Simmons stated. "That is why you and I look so very much the same. Are we the same species, *Homo sapiens*?"

"Yes, each habitable world, at least in our galaxy, is populated with the same intelligent species and our worlds are very similar in size, temperature, composition, density, and atmosphere."

Captain Simmons stood and turned to the wall, her head down in thought. "Is it correct to say the AOW are disgruntled males from The Mother's societies?"

"Yes, either souls who have been reborn to The Mother's world or souls who had knowledge of The Mothers on their native worlds and who attained the 'change' from the AOW."

Captain Simmons looked quizzically at Timeye. "What do you mean by the 'change'?"

Timeye looked directly at Captain Simmons and said, "For all of us, and a few of The Mother's people, we will succumb or die due to the second law of thermodynamics or 'arrow of time.' At the time of what you call the ITP or Big Bang, disorder, or entropy, was at a very, very low state. Afterward, disorder exploded and has continued to increase, in the universe as a whole at least. Locally, conditions created by The Mothers can achieve local ordering or a reduction in entropy, though in the universe it always increases. Those who The Mothers change will not experience progressive and gradual cellular death."

"I want to make sure I understand you," Captain Simmons said. "Those changed are immortal?"

"Yes, though that is not the final state, but their 'wear out' is delayed indefinitely," Timeye said.

"I have more questions, but for now, please rest. Perhaps tomorrow you will help us adjust our course to arrive sooner at The Mother's world?"

"Of course, we want to arrive as soon as possible as well. One last thing, however. You should not let those on your world know about our presence here or of any further course changes. Your world still has AOW spies, and they have the technology to read thoughts within a limited distance."

"OK, you and your people and this device or map will remain secret to the Mission Control people. My crew will know of your presence and origin, but no one outside my senior staff will know of the map device. I must say, however, I am concerned about the presence of spies on the Earth," she answered with a serious look and tone.

CHAPTER 50

Flirting, Pheromones, and Efficiency

"CAPTAIN, STUART, PETER, traversing the wrinkles is not as simple as how you did it the first time when you evaded the AOW. You got lucky and were able to make a simple transit. Each wrinkle has stars on the other side. Some wrinkles are larger, taller than others, so crossing through them will result in a greater distance savings. Lastly, some wrinkles require entry at specific angles and velocities. The velocity depends on the mass of the space vehicle that will pass through the wrinkle and the safe exiting velocity that is required upon passing through. Our device can calculate these parameters as well as an overall optimized course. We will need to know a great deal about your spacecraft in terms of its mass, center of gravity, propulsion limitations, and more. Who can give us this type of information?" Timeye asked.

"That would be me," Peter answered. "Let's go down to the navigation and guidance center. It will be easier to sit and bring up this information down there. I believe it may be advantageous for us to build an interface between our navigation and guidance computer and your device. I'll get Dieter Kratz and Stephanie, our senior software and guidance engineers, to help us out. Sounds like this could take at least a few days?"

"Yes, you are correct," Timeye answered. "We will want to run the entire flight plan through simulations before we execute it for real."

* * *

"How am I supposed to determine space coordinates when these symbols are unfamiliar to me?" Dieter asked.

"I am building a table of my outputs and will be converting them to your numerals," Timeye answered.

"OK, well, let me know when you have that completed," Dieter said. "You know it is interesting how your civilization produced base ten as your numbers just like us. I suppose it results from having ten fingers, just like we do?"

"Of course it is, Dieter," Stephanie answered.

Dieter snorted. "Well, thank God for consistent evolution!"

"I find it interesting," Timeye said, "that you still refer to The Mothers by the male-centric term God."

"We never had a concept of a female deity," Stephanie said, "at least not in modern religions. Of course, *deity* is now gender-neutral in the new age spiritual theology."

"We do not think of The Mothers as deities," Timeye said. "We respect The Mothers, but we do not worship them. I suppose for outsiders, our reverence for The Mothers might seem like worship, and it may be forgiven to think of them as deities."

"Who are 'The Ones'?" Dieter asked.

Stephanie stopped her work and looked at Timeye intently, awaiting his answer.

"The Ones are those who progress beyond The Mothers' world level. They have overcome and experienced all that the physical universe has to throw at them," Timeye explained. "Did I speak correctly by using that idiom, 'throw at them'?"

"Yes, I think so, Timeye," Dieter said. "Though English is not my first language."

"I suppose your native tongue is German. I can speak German if you prefer it," Timeye said.

"Not necessary. I went to the States for my higher education," Dieter answered. "I am fluent in English."

"Well, you are fluent, but you still have that adorable German accent." Stephanie laughed as she said that.

"Are you flirting with Dieter?" Timeye asked.

"Flirting? No, of course not! I'm just teasing him a bit," Stephanie answered somewhat defensively, with a flush on her face, which Timeye noticed.

"She likes you, Dieter. Do you like her as well?" Timeye asked innocently.

"What? Well, I 'like' her," Dieter said, sounding a bit flustered, "but we are not romantic."

"I take it that none of you can detect pheromones. We can," Timeye observed. "Mating is much easier for us because of this ability with which we evolved."

Dieter and Stephanie looked at each other, clearly embarrassed. "I can help your crew to determine who is romantically thinking about who if you wish. In the situation you are in, needing to breed prodigiously, I think it would be helpful," Timeye said without embarrassment.

Stephanie answered quickly, "Timeye, it would be best if you did not provide that service to us. We have our own methods of determining those to whom we are attracted."

"Yes, you call it dating or courting, am I correct?" Timeye asked. "It is very inefficient and leads to feelings being hurt because of misunderstandings—"

Stephanie interrupted, "I suppose you are correct. Perhaps it would be best if we get back to our work. But I do want to know more about The Ones. Maybe we can discuss that later?" Stephanie asked, clearly trying to change the subject.

CHAPTER 51

BLUFs, Acceptable Risks, and Potty Breaks

"WHAT IS THE BLUF, ladies and gentlemen?" Captain Simmons asked the assembled personnel.

"BLUF, what is that?" Timeye asked.

"Bottom line up front," Peter answered.

Timeye still looked confused. "I am still unsure of the meaning. Please elaborate."

"It means that we eliminate the potentially long discussion of how you, Dieter, and Stephanie arrived at the new course and simply state what the conclusions are at the start of our discussion," the captain explained.

"Stephanie is your guidance and navigation expert," Timeye said. "She should state our bottom line up front."

"We can get to The Mother's world in eight years by passing through sixteen wrinkles," Stephanie summarized. "Obviously, this saves us hundreds of years."

"I also want to state that this is a high-risk course. The probability of success is only 98 percent with 95 percent confidence," Timeye stated. "We could collide with any number of celestial bodies upon exiting four of the wrinkles, which have not been suitably mapped for objects just beyond our exit points."

"I appreciate your concern, Timeye, but this sounds like it is an acceptable risk," Captain Simmons answered.

"We would not consider it acceptable, captain. We require 99 percent probability of success with 99 percent confidence when we travel. The level of risk here is outside of our decision-acceptance parameters," Timeye said solemnly.

Captain Simmons sighed and said, "Yes, well, we are a mission of exploration. We have accepted worse odds in the mission to this point."

Peter looked perplexed and asked, "Timeye, you seem to be quite risk-averse. Why is that?"

Timeye looked at Peter and the captain and said, "We view all life as unique and want to allow every form of life the chance to live to the full measure of their creation."

"That is an odd phrase," Peter said.

"We do not believe in accepting risks that wantonly flout the continuation of life and the progression opportunities contained therein," Timeye explained. "I chose this 'odd phrase' from texts of some of your twentieth-century religious cults. I thought it was appropriate and would help you understand my concerns."

"Even so, we are willing to accept the risk, Timeye. Well, I should ask each of you first before I conclude that. Everyone, is this acceptable?" Captain Simmons looked from person to person. Each person stated their acceptance of the risk.

"I have one last question," the captain said. "I believe you three developed this course and the probability of success from Monte Carlo simulations?"

Peter answered, "Yes, I reviewed the one thousand simulations and went through the interface they developed. Stuart also did the same."

The captain looked satisfied and asked, "OK, so we have a new course. When do we implement it?"

"At 10:00 a.m. tomorrow," Stuart answered.

Captain Simmons smiled and said, "Peter, let the crew know of this plan. Tell everyone to go potty before we leave on this new journey."

Bewildered, Timeye asked, "What does she mean 'go potty'?" They all laughed and smiled at him.

Stephanie explained, "It means that everyone needs to be ready for the new journey and that they should go to the bathroom before we leave."

"I am still in a state of confusion," Timeye said, looking concerned as everyone laughed again.

CHAPTER 52

Time's Up!

KANAAN SAT AT a small table with two men standing in front of him; behind them, three security guards waited.

"My lord, the report is that the suspected telepaths were, in fact, not that. We will keep looking," Kush said.

"This cannot go on forever. One telepath will likely reach The Mother's world. A second one, or more, means that The Mothers can form more worlds. This will put our plans at risk!" Kanaan shouted.

"Failure is not tolerated! Take him to the air lock," Esam, the second man, shouted.

The security guards grabbed Kush, who began protesting, "My lord, I beg you, spare me. I have been loyal for thousands of years!"

"Wait! If I execute every failure, it won't be long until we will be down to you and me, Esam, and what will I do to you then?" Kanaan said with a stifled laugh, looking Esam in the eye. "Kush, go back to the Earth, gather up our people, prepare them to evacuate to this location. Wait for my direct order to leave, however."

Kanaan and Esam both sat down as everyone else left.

"Are you really going to do it, execute the protocol?" Esam asked Kanaan.

"I've received word that there is another world where it is known there are telepaths. My lord does not need this world anymore, and he does not want it to be reclaimed by The Mother," Kanaan explained. "We have learned that it is her plan to reclaim this world and give the inhabitants the opportunity to accept The Mother's ways. My hands are tied. He has not ordered me to do it, but it is what he wants done. He wants to have 'plausible deniability' among the Quorum of Elders. My star is tied to his. My success is his success. But I must always protect him, even from himself," Kanaan said.

CHAPTER 53

The Protocol

DEEP UNDERNEATH THE Sea of Tranquility, a giant fusion reactor began operating for the first time ever. The enormous heat produced by the reactor vaporized a giant underground reservoir of water that the AOW had created under the surface of the moon. The vaporized water pushed upward on the crust above it, opening cracks on the surface. The creation of these cracks caused a flurry of moonquakes as large pieces of the crust uplifted and broke apart. With the crust broken, the liquid outer core of the moon began to expand.

A second reservoir of water was heated to vapor by an even deeper second fusion reactor. The water vapor in the second reservoir was contained until the moon's expanding outer core of nickel and iron broke the reservoir. The superheated water vapor rapidly escaped the reservoir in a gigantic explosion. This caused large sections of lunar crust to be ejected from the surface. The moon's weak gravity could not retain these pieces, which had enough velocity to escape its gravitational grasp. Several of these gigantic chunks headed toward the Earth. As they got closer, the Earth's gravity pulled on them, increasing their velocity. Within minutes, over thirty pieces of lunar crust entered the Earth's atmosphere. The sonic booms of the entering chunks were the first thing people noticed. Seconds later, the first of the chunks hit land near the abandoned city of Moscow. The collision was so violent that the tilt of the Earth shifted six degrees within seconds. The expanding shock wave spread worldwide within two hours, flattening cities and even breaking away pieces of mountains. Large sections of the atmosphere ignited from the enormous heat that followed the shock wave, causing the oceans to boil.

* * *

Edward, Terra's grandfather, was sitting in his family room in the suburbs of Denver with his wife while his adult daughter, Anne, put her two children to bed in the spare bedroom.

As he sat there, a blinding light engulfed the whole house, inside and out. This was followed immediately by the loudest noise Edward had ever heard. His eardrums burst, and he was shredded by semi vaporized glass from the windows. This was the last thing he remembered before losing consciousness.

Edward awoke hours later. It was dark. Whether it was because it was night or whether the day had been turned dark by debris covering up the sun, he could not say.

Edward slowly remembered the last things he saw before he lost consciousness, the light and the boom.

He struggled to breathe. The air smelled like ash, and it left him choking as he drew breaths.

Where are Kelly, Anne, and his grandchildren? Edward thought.

He panicked and started crawling around in the darkness. He soon realized it was futile. He would have to wait until dawn. The hours seemed to drag on and on. He must have eventually slept because when he awoke again, it was light though it seemed as if the sun was barely working. It was more like twilight than day.

He did not know it, but he had survived a near-extinction-level event. He was one of the lucky ones, assuming that surviving this event would ever be considered "lucky." He did not know it then, but over 99.999 percent of human life had been extinguished.

CHAPTER 54

AOW Spaceship *Intrepid*

KANAAN COULD NOT watch, at least not at first. As the first piece of the moon hit the Earth, Kanaan turned away. He shook violently, as if he were having a seizure. Kush moved over to him to help him.

"My lord, are you OK?" he asked.

"I will be," Kanaan said. "Surely you must feel great sorrow at this result also?"

"I take no joy in this, But it furthers the aims of our lords," Kush answered.

The spaceship they were on, the AOW *Intrepid*, was far enough away that they were safe from any ejected mass from either the moon or the Earth but still close enough that the enormous fireball overwhelmed those watching from the observation bay, forcing them to shield their eyes. Each of those looking on felt something akin to a great loss as they watched the end of the world on which they had been born.

"My Lord, what of the off-world colonies, Mars and Europa?" Kush asked.

"We will leave them be. They will all die without support from the Earth."

"But they will send word to the *Calypso* should it still exist, my lord."

"Let them. They will then know we mean business and we do not fool around!" Kanaan answered harshly.

CHAPTER 55

Musk Colony, Mars

TARIQ SHEPHERD SPOKE again into the EPC comm. "Please repeat, Earth, please repeat. We experienced a comm failure of some kind." Tariq waited, but no reply appeared in the EPC tubes.

He tapped his wrist comm device and spoke to the satellite room across the hall from him. "Shandra, can you raise shuttle services at O'Rourke on your light speed comms? We are experiencing an EPC issue."

"I would if I could, Tariq. My last call with satellite services in Baltimore ended mid message. We've got some issue also," she answered.

Tariq started to walk across the hallway to talk to Shandra but ran into her as she was on her way to talk to him.

"What on Mars is going on?" Tariq asked.

"I don't know, but whatever it is, it is affecting Earth and O'Rourke comms. We need to get to the observatory room. Maybe there was a solar flare?"

"Well, that wouldn't affect the EPC," Tariq said with a worried tone.

In the observatory, they found a large group of people assembled at the front of the room. Shandra started to say, "Can any of you tell me why light speed and EPC comms are down between here, O'Rourke, and Baltimore?"

An older man, Petra Thorsen, turned toward her and said, "The Earth is gone. The moon experienced massive quakes, and pieces of lunar crust hit the Earth. It's all gone, lunar bases one and two, O'Rourke, and the Earth. Gone!" he said loudly but with shaking in his voice.

"What do you mean gone? Ain't no way. You hopped up on something, man?" Tariq said in disbelief, but cold fear was growing in his gut.

"They're destroyed," Petra said. "Massive shock waves from the impacts have destroyed all life on Earth, the moon, and O'Rourke."

Tariq felt dizzy, Sandra, dazed and frightened. In unison, Shandra and Tariq's knees became wobbly as they leaned against the wall and slid down to the ground, holding on to each other as they both burst into tears.

CHAPTER 56

FBS *Lightfoot*

KWAME WAS GLUED to the observation panel, watching as explosion after explosion hit the Earth. He was on his way to Mars for a well-earned vacation with his family. All around him, crew members were talking, some sobbing, and some just silent as they watched the Earth absorb hit after hit with lunar crust.

The ship's commanding officer, Jared DeSoto, entered the observation lounge. "Can I have your attention, please? I have some information, please your attention.

"We've lost comms with the Earth, lunar colonies, and O'Rourke. We surmise that this is a result of what we are all observing, the impact of lunar crust on the Earth and massive quakes on the moon. O'Rourke has been affected also, probably by debris. All we can assume, for the moment, based on what we have observed, is that they have been completely destroyed. We received a short, incomplete message from lunar base one that said they were experiencing massive moon quakes. Those quakes broke free lunar crust that impacted the Earth. Our telescopes show that the Earth sustained up to thirty impacts from the crust pieces. We are proceeding on to the Musk Mars colony. We will let you know more as soon as we know more ourselves."

Kwame knew more than anyone in the solar system about the situation and who was behind the destruction. He knew he had to get to Scott Pettis at the Musk colony ASAP. Pettis was a senior adviser to President Perrin and had been "read in" on the *Calypso* and the AOW. They had to ascertain if anyone on Earth was still alive.

The secret Mission Control room at Groom Lake was a sealed enclosure, meant to survive catastrophes. Communication with the *Calypso* via EPC was available there and nowhere else. The *Calypso* needed to know about this event. Mission Control no longer received positional updates from the *Calypso* to keep them safe from the AOW, who might be listening. EPC was the only method of feasible communication now, and matched pairs of particles in the EPC only existed at Mission Control and on board the *Calypso*. Kwame also needed to talk to a planetary specialist. If the Earth's atmosphere reformed after the debris settled, there might be other survivors. They needed to start contacting them ASAP.

CHAPTER 57

Several Days Later

EDWARD SINGER MOVED through the debris of what was once his home in the Denver suburb. His wife, Kelly, and adult daughter Anne with her two children were with him. They were looking for anything that could be useful.

Useful? What does that mean in this situation? Could we even survive in this hellish world? Would we WANT *to survive? We still don't know what really happened. A nuclear war? With whom? Nuclear weapons still existed, but the old adversaries whom they were built to protect against have been dead for at least two generations now,* Edward thought.

"I found some photo albums, Edward," Kelly said. "I know we really want to find the baby formula and other baby items. I'll keep looking."

"Mom, the kids need to rest. I'm going back to the warehouse," Anne said.

"Wait!" Ed shouted. "I don't want you to walk back alone. There are hungry and dangerous people out there."

Kelly stopped looking and stared at the ground and said, "Ed, I think we can go back. We've looked through this twice now. We'll have to find baby formula somewhere else."

"OK, let's go," Ed said, exhausted.

* * *

"If I am hearing you correctly, you are saying there could be survivors—not all regions of the Earth experienced a sudden loss of an atmosphere. The atmosphere has reformed over the Earth," Scott Pettis said, summarizing what Dr. Ginger Kimball had explained to the assembled group.

"Yes, the initial assumption that the atmosphere was blown away or ignited was not completely accurate. Some mountain valleys and more enclosed areas, like the Grand Canyon, for instance, kept an atmosphere throughout the impacts and the ensuing shock waves. There are survivors—few, I would guess, but some. Your facility in Nevada probably has survivors, based on the facility specifications that I have reviewed. Their shock isolators and dampeners mitigated the ground motion, and the atmospheric sustainment system could have maintained a survivable atmospheric pressure," Dr. Kimball explained.

"Gus, have we heard anything from Earth over light speed comms?" Kwame asked the base director, Gus Johnson.

"We have picked up a number of SOS calls from dozens of locations across at least three continents."

"Anything from Groom Lake Control?" Kwame asked hopefully.

"We did get some comms from Nevada, but we don't know if it was Groom Lake. The communication was broken and garbled. We heard a name, Bruce Henderson. Is he somebody in your control center?" Gus asked.

"Bruce Henderson, yes, he is the control room manager! What did he say?" Kwame asked excitedly.

"He was trying to get hold of emergency services, then the connection was lost."

"Can we try back again on the same frequency?" Kwame asked.

"Sure, there is no guarantee he will hear or answer. We'll identify ourselves. You can try speaking directly to him—maybe he will catch that, and he will keep up a regular conversation with us on a schedule. I do have to ask, though, what can we do for him anyway?" Gus asked seriously.

"We're going to go rescue him and the team at Groom Lake, and we're going to take the EPC comm system so we can keep in contact with the *Calypso*," Kwame answered excitedly.

"Do you have Earth-qualified landing shuttles?" Kwame asked.

"Yes yes, we do, but I would need authorization from the FBS to land it—" Gus answered but caught himself and added, "Well, I guess that doesn't really matter anymore. At any rate, Mr. Pettis could provide the authority, I assume." Gus smiled slightly.

CHAPTER 58

Prioritizing the Very Long Term

"WHAT CAN YOU tell us about the debris layering, Captain DeSoto?" Kwame asked.

"We've identified six orbital pieces of lunar crust, size sixty meters to four hundred meters, irregular in shape. We've given you the longest dimension, FYI. We don't see any difficulty establishing a safe orbit," Captain DeSoto summarized.

"Good, now let's talk about the landing priority list. Mr. Pettis is our governmental survivor as of now. He is authorizing that you land your first shuttle at Groom Lake and the other shuttle at the presidential survival bunker in Geneva. At that point, we want to sample two urban areas for survivor estimates. We've chosen Sydney and New York. We will then reassess and prioritize the next landings. For your information, the overall FBS priority is species continuation outside the solar system. Species continuation within the solar system is only possible if our species continues on Earth, which is deemed, currently, to be unlikely. To that end, the *Calypso* mission has priority. Any questions, captain?" Kwame finished.

"I'm rooting for 'species continuation' on Earth," Captain DeSoto said dryly.

"Yes, captain, we all are," Kwame said somberly.

CHAPTER 59

Groom Lake Rescue

"*BUZZ ALDRIN*, COME in. I say again, *Buzz Aldrin*, come in," Sarah Radle said into the comm set from the comm room on the FBS *Lightfoot*.

"*Buzz . . . Ald . . .* ere. We are experiencing significant comm distortion . . . rom . . . ck debris in . . . atmosphere, over," came the response from the *Buzz Aldrin*, in descent to Groom Lake.

"Copy that, *Buzz Aldrin*. We are switching to a routing through a network satellite still in service. This bird is stationary over your LZ. We hope for better comms now, over," Sarah answered from the *Lightfoot*.

"Seems better now, *Lightfoot*," *Buzz Aldrin* responded back.

"We don't expect video to be possible. "Position, status, what do you see?" Sarah asked.

"Descending through fourteen kilometers. Bumpy. Deployed high-altitude primary parachute. Parachute shredded by atmospheric debris, switching to rotors now. No visual on the surface yet, cloud and debris cover extensive. Descending through two kilometers, clouds breaking up, I have a visual on the surface. No surface structures visible. I can see the runways, possibly intact.

Touchdown in twelve seconds . . . touchdown. We are sampling air and soil on the planetary sampling system. Stand by *Lightfoot*."

"Copy that, *Buzz Aldrin*," Sarah answered.

"We've got atmospheric pressure equivalent to three thousand meters. Original Earth specs show that the altitude here was six hundred meters above mean sea level, meaning we see either a partial loss in atmosphere or a significant change in mean sea level or a combination. The temperature is freezing, zero degrees Celsius. Oxygen content as expected for Earth at three thousand meters. Toxic content, particulate, and gaseous species within safe limits," the *Buzz Aldrin* answered.

"Copy that, *Buzz Aldrin*."

"We will commence EVA as soon as we can switch to environmental suits, estimate thirty minutes."

Rodney Langford and Donovan Willis unfolded a base map to help them locate the entrance to the *Calypso* control center.

"Without topside features like buildings, this is going to be a bit of a guess," Donovan said.

"I know, but we can geolocate from the surviving satellites, and the elevator shaft should be . . . here, right here!" Rodney shouted in triumph.

"Yes, looks like we are going to be digging for a while. *Buzz Aldrin*, can you get Becky and Tristan to get the excavator out of the cargo bay?" Donovan asked over his suit comm.

“Copy that, Donovan,” the *Buzz Aldrin* answered.

“I think we have located the blast door,” Becky said to Rodney, who was still standing on the edge of the hole that had once contained a functioning elevator. “There is a door communication voice and data pad. I’m going to try the voice comm and see if anybody inside answers.”

“*Calypso* control center, this is Becky Telford from the FBS *Lightfoot* out of Musk Mars colony. Are there survivors? Over.”

A full minute went by while Becky repeated the same question before an answer.

“Becky, so happy to hear a voice, even if it is a Martian voice!” came the reply.

“Copy that, control center. What is your status? Over,” Becky answered back.

“We have twenty-seven survivors. What is it like topside?”

“Not so good, control center. We’re here to take you to Mars.”

“Never been there. I guess the world had to end for me to get there. Not how I wanted it,” the control center answered.

“I can understand that completely,” Becky answered.

CHAPTER 60

FBS *Calypso*

CAPTAIN SIMMONS STOOD behind Sam as he read the message to the assembled group that included Stuart, Peter, Gunner, Timeye, Simi, and Terra. Each of them sat down in the chairs around the conference room table to take a few minutes to absorb their grief. No one spoke for several minutes. Terra sobbed silently as she realized her grandparents and cousins were dead.

Captain Simmons thought, *Everyone knew this mission was a one-way trip. Maybe this helped deal with the otherwise difficult news.*

"I will make this known to the crew. There is no reason to keep this secret. It looks like our support from Mission Control will be more limited. At least we still have EPC, and we will know some of the people on the other end of the communications we will receive going forward," Captain Simmons said.

"Captain, I think we need to talk about the hibernation schedule. We are currently six years from The Mother's world, and we are in a conflict with an alien and superior race, at least technologically. Would it not make sense, from a mission success standpoint, to have the current crew stay awake for the duration of the voyage?" Peter asked.

"Maybe, but what of our compatriots from Caldonia? Timeye, Simi, would your people want to endure a voyage that long?"

"We never expected to be put into your hibernation protocol. Our lifespan is normally seven or eight hundred of your years, anyway. Six years for you is only one or two years for us. We could help you extend your lifespan. Would you like us to do that?" Timeye asked.

"Yes," came answers from all around the table.

"So we shall then. I understand that Terra is a proto-Mother and has an unlimited lifespan," Timeye said.

"I do?" asked Terra.

"Yes, you are an advanced soul, which is why you are a proto. You have the gifts."

Simi smiled and said, "Yes, Gunner, I would like that!"

"What?" Gunner responded, confused.

Terra rolled her eyes at Gunner. Captain Simmons looked at Terra and Gunner, also confused.

"Gunner wants to be intimate with me," Simi responded.

"I didn't say anything! Oh yes, the pheromone thing again. Terra, you're not reading my thoughts, are you? Simi, is it really that easy on your planet to get laid?" Gunner asked with a smile.

"Laid? Timeye, what does he mean?" Simi asked Timeye in her tongue. Timeye explained; then Simi smiled at Gunner.

“People, we are getting off subject here. Everyone, put on your professional hats! There is a child with us!” Captain Simmons shouted in exasperation as she stared down Gunner, whose face was now turning red.

“I’m almost fifteen!” Terra said indignantly.

Well, at least this sexual innuendo lightened an otherwise somber moment, Captain Simmons thought as everyone left the conference room.

CHAPTER 61

The World of The Mothers

EVE OF EDEN moved slowly down the hallway, sobbing silently as she entered her private chambers. Elotim got to her just as she fell into his arms and passed out. He carried her over to her bed and laid her down. Within a minute, she opened her eyes and stared up at the ceiling.

She remained motionless for a couple of minutes before saying, "Kaanon has betrayed The Ones. He allied himself to the AOW. They were searching on Earth for more who had what they call the gift. When they found none, the Council of Elders ordered the Earth destroyed. There are survivors on Earth—few, but some. Some of our children remain on Mars and Europa, but they will die eventually without the Earth. We must save them and bring them to the new world."

Elotim sighed. "Ever since we were forced to leave because of the comets, we have had to rely on those we left behind. Those who were not born to the level of our world. Those who we could not trust completely. Now, we learn that all our work is gone. I have no words for the sadness I feel. My child, when we left, we left under the direction of The Ones. They know the end and the beginning and all things in between. They knew we would have to deviate from the protocol and nurture that world through others. They knew that this would cause great violence, selfishness, and suffering. Yet they allowed the Earth to continue. There has been great progress made by many souls in these difficult circumstances. When we bring survivors to a world near ours, we can guide what remains into a strong, just, and peaceful world."

Elotim thought for a moment and then said, "To bring them to the world near ours, we will need a large spacecraft. We have not needed spacecraft for so long that what we have will not hold a great number. We will need to build landing craft again and a large interstellar vessel. We must build these ships. We cannot wait for one tick."

"Father, we could use the *Calypso* as the interstellar vessel. We could greatly expand the number of hibernation berths. We could build landing craft for the rescues much more easily than we could build a starship. This is the only way to proceed, don't you agree?" Eve asked.

Elotim thought for a moment and then nodded his head slightly.

"I will need to tell Terra to contact the Mars and Europa colonies. They need to know a rescue is on the way, but it will take more than a dozen years. They must plan accordingly," Eve said with determination.

* * *

Terra was engrossed in her studies. As the hours passed, she became weary and eventually laid her head down on her tablet and fell asleep. As she slept, she dreamed.

The Mother stood in front of her on a broken landscape. Dark clouds of dust and debris nearly blocked the sun. The Mother beckoned for her to follow. Terra walked behind The Mother. Eventually, they stopped next to a partially destroyed building. They entered and found people gathered around a fire, trying to stay warm. Two small children were crying in the arms of adults. Terra surmised these people were hungry and desperate. In one corner of the room, telecommunication equipment was set up, and a man sat in a chair next to the tables with the equipment on it. He was bent over a small speaker, listening intently. Terra heard the call sign of the voice, "Mars ComCent."

As she listened, the man replied, saying, "We have one hundred and forty souls at our location. We know of three more groups near ours, about the same size, over."

"Copy that, Denver," Mars ComCent responded. "Help is on the way, but it is more than thirty days out. Hang in there."

"Terra, these people are survivors on the Earth," said Eve of Eden, The Mother. "They live in your hometown, Denver. They will be taken to Mars, but the capabilities of the Mars colonies will be stretched to the breaking point to keep all these people alive. When the *Calypso* reaches my world, your ship will be retrofitted so that twenty-five thousand people can be put into hibernation. Some of your people, and some of those in my world, will return and rescue those who survived the destruction wrought by the AOW. They will bring them here, and you shall be The Mother of this new world. There are modifications to the *Calypso*, which can be made now prior to your arrival, which are necessary for the rescue to be possible. I will put the plans for these modifications into your central AI. Your people will need to begin these modifications immediately."

CHAPTER 62

Bonding

GUNNER ROLLED OVER onto his back and looked over at Simi who smiled widely.

"I've got to get going. We have full days now. Drilling out rock and sealing the walls of the new chambers below the reactor spaces," he said.

"Will you be going back to your solar system once we reach The Mother's world?" she asked.

"Yes, Captain Simmons wants me on her return crew," Gunner said. "I can say no if you want me to."

"No, I will go with you," Simi said. "I can help heal those who suffered so much."

Simi, who was completely naked, stood up and turned to face him, smiling. "I love you, Gunner. I will always be with you no matter where you go."

"I love you too, blondie, you're my honey."

"We are soulmated," Simi responded. "That is the proper designation, my love."

Gunner looked at her, his eyes full of love. "I know, I know. I'm still learning about your culture and how to translate your words into my language."

"I have a full schedule of counseling today. I won't be home until 7:00 pm. I am with child, Gunner," she added with a smile. Gunner's eyes widened, and he was silent for a moment. "Does this not please you, Gunner?"

"No, it does, it does. I'm j-just . . .," Gunner stammered.

"Afraid? Yes, I sensed that," she said. "There is nothing to be afraid of. I am very good with children, and I will still be a good wife to you though I will get fat."

"Oh, don't worry about that. I find pregnant women strangely sexy, anyway."

"You find a great many women sexy, my dear," Simi said, without a hint of jealousy.

"I suppose I can't deny that since you cannot be lied to," he said with a guilty smile. "I must be off now." He stood up and walked toward the door.

Elsewhere, Captain Simmons was in a meeting with the Hamm sisters, Jackie, Jenny, and Jill. The sisters were identical triplets. This odd genetic anomaly occurred more often in the twenty-fourth century with selective embryo selection. The sisters previously ran the world's largest construction company, Triple J, which had been among the myriad of contractors that built the *Calypso*. Their job now was to create the two large new caverns for extra hibernation pods. The sisters had been among over one thousand people in hibernation that had won a lottery for their spot on the *Calypso*.

"How are we going to remove that amount of material to the topside?" Jackie asked.

Jill answered, "We'll have to do it through the secondary ventilation shaft."

"That means there won't be a secondary ventilation shaft, which was put there for a reason," Jackie countered.

Jenny was exasperated. "Will you two stop arguing please!"

"When you three have a unified position, come find me and brief me on what you decided," Captain Simmons said in obvious frustration, as she walked out of the engineering conference room.

Timeye was in the hallway, waiting for the captain. "I sense frustration. The Hamm sisters again?"

Sandra smiled at him and said, "You probably don't have to be a Caldonian to nonverbally sense my frustrations."

"But it helps," Timeye answered. Captain Simmons smiled, leaned toward Timeye, and kissed him on the lips.

"I am warming to the concept of the kiss," Timeye said softly as he leaned toward the captain and kissed her back.

CHAPTER 63

Digging

THE DRILLING AND careful blasting of the interior of the asteroid was a slow process, primarily due to the tedious effort required to remove the rock to the surface. The secondary ventilation shaft had been converted to a conveyor belt designed to move the removed rock to the surface. Then the rock had to be melted in a blast furnace constructed there. The molten iron ore was removed and mixed with carbon and other alloying elements, taken from ship stores, to make the desired steel alloy for the giant new hibernation bay. This bay would be below the reactor space, which was the deepest area in the asteroid until now. The hibernation pods themselves would be built by The Mother's people and would be installed into the new hibernation bay upon their arrival at The Mother's world.

The Triple J sisters oversaw the entire project. They had a crew of forty-five excavators, thirty blast furnace workers, and another thirty mill workers who produced the steel panels and structural beams. At a future time, groups of steel framers would assemble the structural beams into the rock cavity. Riveters would then hang steel panels to the beams. Welders would weld each panel joint and seal the joints with high-strength epoxy to make an airtight compartment.

Many of the workers were people from the hibernation pods. They were awakened and briefed on the project and on the status of the Earth and all other voyage developments. The newly awakened workers needed living quarters. Some crew quarters were made vacant by some of the existing original crew going into hibernation, but this did not alleviate the shortage of crew quarters. More individuals now shared each crew quarter. This made for a more cramped living arrangement, but the crew made the best of the situation.

CHAPTER 64

Terra and Hatcher

TERRA AND HATCHER celebrated their twentieth birthdays just days apart from each other in July, more than ten years after they had left Earth. Hatcher was a handsome young man, over six feet one inches tall with brownish-blond hair and blue eyes. Terra was, well, beautiful. She was five feet six inches tall with dark hair and green eyes. She was soft-spoken and kind to everyone. Her special abilities had made her an oddity around the other young people her age. The adults, those older than her, understood better who she was and seemed to accept her more easily than those her age.

Hatcher was a fully qualified ship technician and had learned to fly the landing shuttles, which they would need when they arrived at The Mother's world. Terra was engaged in her physics and engineering studies and was close to earning a master's degree in both. She was also training to be a ship's officer. She had switched her career interest from communications engineer to executive officer. This meant she would have more years of training, working in various departments on the ship. Now she was an intern in the reactor space.

Terra's parents had big news. They were having a baby—twins, in fact. Multiple births were encouraged since the fate of their new world was primarily dependent on reaching a critical population level that could survive disease and other hardships they would encounter on their new world.

Terra and Hatcher were to be married once they both reached their twentieth birthdays, which would be about two years before their arrival on The Mother's world. Earlier marriages were also the norm now as well. Life on their new world was going to be tough in so many ways, ways that human beings had not experienced for hundreds of years. Terra and Hatcher planned to stay on the new world while the *Calypso* returned to the solar system to pick up survivors from Mars, Europa, and Earth. The new world needed many building projects in anticipation of the many new settlers who were survivors from the Earth's solar system. Terra was unsure of what her new duties would be as "The Mother" of the new world. She assumed she would learn them when she met with the original Mother and her "colleagues" on their planet.

"Mom, how do I look?" Terra asked as she finished putting on her wedding dress.

"You look beautiful, honey," Riley answered.

"I'm so nervous I'm not sure I'm going to keep myself together. I have looked forward to this day for years, but now I just want it to be over. I want to be married to

Hatcher, nothing else matters," she said, almost out of breath with excitement and anxiety.

"Well, that is going to be very soon now. I see that the captain has arrived," Riley said.

* * *

"I'd like to welcome all who have gathered today to celebrate Terra and Hatcher's binding together. I think this is just about the biggest gathering we've had since before we left the asteroid belt. I know not everyone could be here, but we still must keep the ship running. I'm told this will be the number one watched program on ship TV today," Captain Simmons said to laughter from the assembled crew in the hangar bay.

"Terra and Hatcher represent the first couple who will be bound together from among our crew. There are a few others who could have taken this step." Captain Simmons paused as she said this and looked at Gunner, Simi, and Timeye, who all turned red as she said this, to more laughter.

"I know both of these young people very well. Terra and Hatcher were both ten-year-old children when this all started. I can't help but continue seeing them this way, even as they stand before me all grown up. Well, maybe Hatcher has a way to go. I don't think he really needs to shave yet!" Captain Simmons smiled and looked at a red-faced Hatcher.

"Well, let's get to the best part of being a captain in the air and space force, shall we? Terra and Hatcher, do you enter into this binding together of your own accord, without coercion and with a willingness to love, support, and sustain each other through all the years of your lives, through good times and bad, in sickness and in health? Hatcher?" Captain Simmons said as she turned to him.

"Yes, I do," Hatcher said as he looked Terra directly in the eyes.

"Terra, do you?" Captain Simmons said as she turned to Terra, who was crying softly.

"Yes, I do," Terra answered.

"If there are no objections from those here assembled, I grant you, Hatcher Newsom and Terra Nova Winstead, a legal binding together as husband and wife by the powers and rights given to me by the air and space force of the Free Bloc States of the planet Earth. Let's party, everyone!" Captain Simmons said as everyone clapped for the young couple.

The partying continued for three hours. Terra and Hatcher danced the first dance together in front of the assemblage, as was Earth custom. By the time the end of the evening shift came, at 10:00 p.m., everyone was tired and ready to go to their quarters, not the least of which included Terra and Hatcher who seemed to have danced with everyone there.

* * *

Even though they were married, Terra and Hatcher still lived in the quarters of Terra's parents. This was due to the lack of accommodations given the influx of those who were working on the ship modifications.

Terra and Hatcher had just sat down to dinner with Terra's parents. The dinner conversations now seemed to revolve around the conclusion of their voyage and what new and interesting things they would encounter on The Mother's world and then on their new world. So far, they had surprisingly little information about The Mother's world. They knew the people there looked like people from Earth; but they didn't really know what they would be like, how they would act, what customs they had, the state of their technology, and many other things. Terra was anxious about all of this as she saw herself as the one who would interact with The Mother the most. She was going to start asking The Mother about what she could expect.

"How old is The Mother?" Hatcher asked.

"She's thousands of years old," Terra responded.

"How can that even be possible?" Riley asked.

"Well, we don't know, but surely their technology is advanced beyond ours," Sam said.

"I thought she didn't look a day older than sixty years old," Riley said. "Not quite middle-aged, in my opinion."

Terra said, "The Caldonians live to be hundreds of years old. With better technology, that could be turned into thousands of years old, but I'm guessing that the key to living that long is something we haven't even thought about yet."

"What is perhaps more interesting than their technology are their customs, lifestyle, and governing style. Things that may be quite different from what we are used to," Sam mused.

"I hope I can learn some of these things before we get there. I must be going. The captain's weekly briefing is in a few minutes," Terra said.

"Sweetheart, if you see Gunner, ask him if Simi can come by tomorrow. She was going to teach me some pain-coping meditation skills for childbirth."

"Sure, Mom," Terra answered as she got up and walked out the door.

CHAPTER 65

The Deceleration Plan

MORE TIME PASSED, and the prospect of reaching The Mother's world and the settlement world drew closer. Captain Simmons had tasked Stephanie, Dieter, and Terra with developing a detailed deceleration plan. At this week's captain's briefing, the three would be presenting the preliminary draft of the plan.

All executive officers and officers in training attended the captain's weekly briefing. As required, individual engineers and others attended if they were giving a briefing. Terra attended as an officer in training.

Terra called the meeting to order. "Everyone, your attention please. We need to get started on time today, as we have a very important issue to discuss. Today's meeting will be on our deceleration and orbital plan for our arrival into The Mother's solar system. Stephanie Pokorny and Dieter Kratz are presenting the results of computer simulations on how we will decelerate to achieve a suitable orbit and maintain gravity during the time we will be in orbit around The Mother's planet, the first planet, and then again when we are in orbit around the new settlement planet, planet number two in that system."

Terra turned the time over to Stephanie and Dieter and handed the laser pointer to Stephanie, who now spoke.

"Thank you, Terra. As you are all aware, maintaining a level of gravity on our voyage was vital to maintaining our health en route. We will need to position our ship so that our shuttles can reach the two worlds in a short time frame and return to the ship in a reasonable time to load up cargo and personnel and then unload it all again on the settlement world. No matter which of the three orbits we choose, we will need to begin decelerating beginning at about two years prior to establishing orbit, or six months from now. This is a much shorter time than any mission planner assumed. We can do this for two reasons.

"First, we have not been accelerating for hundreds of years, and our velocity never reached anything close to 4 percent of light speed. Secondly, we plan to use the gravity of the two gas giants in The Mother's solar system and a set of solid rocket motors to achieve 90 percent of our deceleration. We will also be deploying the solar sails once we turn the ship and start our deceleration. The particles streaming from the star Cervantes will be 'felt' by then, and we intend to use every bit of it. The deceleration will start by the end of this year and continue for the remainder of our journey. There will be periods of increased gravity and then periods of weightlessness that we will all have to endure, but we believe we have minimized this.

"As I understand it, the medical staff will be presenting the plan for maintaining cardiovascular endurance and bone and organ health during our deceleration at next week's meeting. The three orbital plans are one, establish a long elliptical orbit around the Cervantes star and The Mother's world with an orbital perigee at The Mother's world; two, establish a standard orbit around The Mother's world; and three, establish an orbit around the settlement world. In the case of option three, the shuttles could travel the distance to The Mother's world in about fourteen days, but it would allow for a more efficient unloading of cargo and personnel. One thing we don't know yet is how, where, and when The Mother's people will need access to the ship for the completion of modifications for the rescue voyage. At this time, all we will be doing is presenting the options. From this point onward in the presentation, Dieter and I will be going through the simulation runs. If some of you don't have an interest in this, it would be a good time to excuse yourself," Stephanie said.

CHAPTER 66

Rotate, Burn, Decelerate

LATER THAT YEAR, in December, the day came for the giant ion propulsion system to be shut down so that the *Calypso* could be rotated one hundred and eighty degrees for the deceleration phase. There would be a period of weightlessness while the attitude control engines mounted on the outside periphery of the asteroid rotated it. Things that were unsecured needed to be secured for this period. Tables and chairs, in most cases, had mounting points on the floors to which they could be secured. Livestock had to be secured in weightless harnesses, the personnel had to wear their magnetic boots, and every loose item had to be stowed away.

"Captain, we have full shutdown of the main engine," Stephanie said through her headphone in the guidance and propulsion control room.

"Copy that, Stephanie," Captain Simmons replied from her command console on the bridge.

"Rotational rockets activating, per program step number two," Stephanie continued to announce. "Rotation completed, checking line of force for ion engine ignition and burn. Captain, we will have some positional tweaking as expected. Attitude control rockets engaged to adjust positioning,"

A few moments went by as the ship was nudged back and forth by the attitude rockets.

"Captain, we have proper positioning. Deploying solar sails, this will take about an hour for the sails to deploy and form up as they pick up particles. I will advise when this is completed. You can watch this on camera number four, though this might be a bit boring,"

"Copy that," Captain Simmons replied.

Captain Simmons and Peter Chan watched as the long steel cable unwound from the giant reel. Everything seemed to be going smoothly until the cable line fully unwound. At that point, the sail packaging was supposed to open and unfurl via spindly arms that were meant to help the sail fully deploy. The sail started to deploy, but one-half did not. The result caused the sail to rotate slowly as the forces on it were unbalanced.

"Captain, we have a problem!" Stephanie said, obviously concerned.

"Copy that, Peter, and I can see the sail is only half deployed."

"Yes, but the real problem is that we are going to start moving in directions that we don't want to move. The ship could become unstable and uncontrollable very quickly. I recommend we cut the sail line right away," Stephanie said.

"Do it," Captain Simmons said without hesitation.

Stephanie interrupted the program execution and commanded the line to be cut explosively at the reel located inside the cavity that housed the sail and tether line. This was done automatically by an explosive charge located on the reel drum. The captain and Peter watched as the line snapped and floated away from the ship, taking the solar sail with it.

"The tether line is severed, captain. Recommend we continue with the program execution and ignite the main engine."

The captain, concerned, asked, "What will that do to our deceleration profile?"

"We will need to reassess, captain."

CHAPTER 67

Slowing a Runaway Rock

"IN SHORT, ALTHOUGH the sail was only going to provide 10 percent of our deceleration, without it, we will not be able to establish an orbit around The Mother's world," Stephanie concluded.

"Does anyone have any ideas on how to slow our ship down more?" Captain Simmons asked.

There was muffled talking in the conference room for a couple of moments before Timeye cleared his throat and began to speak. "Yes, there is a possible method of gaining extra braking . . . Is that the correct word for this context?"

Captain Simmons smiled at him. "Yes, that is the most appropriate word. What are you thinking, Timeye?"

"I understand that you have a little over ten million gallons of water in the coolant tank. I believe we could construct a tea kettle and use steam as thrust to slow us down. The hot water coming out of the reactor could be heated to steam in a boiler. This could provide thrust to slow us down. We would need a nozzle, or better yet a series of nozzles, to be constructed and a throttling system that will allow for control of the steam output."

Stuart asked, "Wouldn't that require most of the water in the coolant reservoir? How would we cool the reactor?"

"We would shut the facility reactor down after vaporizing the water we need for braking thrust. We would save enough water to be able to run the propulsion reactor since we will need to transit between The Mother's world and the settlement world. Pulling 80 percent or so of the coolant water for braking thrust would have to be done within days before our arrival since we would only have a few days of power from the backup batteries and solar panels," Peter explained.

"Peter is correct," Timeye said. "He ran calculations on how much water would be needed to cool the reactor and provide enough thrust for deceleration and when we could eject the steam."

Peter explained, "It's close. We will need to fully charge the batteries, deploy extra solar panels, and develop a plan to shut down everything but the most necessary functions on board."

"Peter, Timeye, I want a second team of engineers and scientists to look over your plan to give me confidence we can do this. Stuart, can you get a group together to go over this plan in detail?" Captain Simmons asked.

"Of course, we'll get right on this."

"Terra, we need to know that our water reservoir can be refilled at The Mother's world," Captain Simmons said. "Specifically, do they have the capability to lift that much mass up to the *Calypso*? I assume they do, but we will need to get enough water up to the *Calypso* so that we can restart the facility reactor ASAP once we arrive. We have people in hibernation, and they will need to stay that way until we get things set up on the settlement world."

CHAPTER 68

Desperate Times on Mars

KWAME RELISHED HIS once-a-week bath. Water had become far more precious than gold. Every drop of urine, humidity in the air, and fluids from the deceased were being recycled. Hygiene was the subject of many jokes in the weeks following the events on the moon as draconian water restrictions were put in place. The whole base seemed to smell like a locker room as everyone ripened up. Water extraction from the polar soil ice was being ramped up, but the demand always seemed to outstrip the supply. The three hundred refugees from Europa and the ship full of refugees from Earth had strained resources. Mars was not a planet humans could live on without massive resources from Earth. Even so, they had to do things outside the box to survive.

Kwame finished his shower, toweled off, and headed to work. He was still acting as the director of the *Calypso* mission though his staff was now just twelve people. The EPC comm device had been retrieved from Earth, and it allowed daily communication with the *Calypso*. They knew the *Calypso* was braking for arrival into the Cervantes system, they also knew about the solar sail failure, and the teapot scheme to slow the ship. The survival of the *Calypso* was now tied to all their survival. A return trip was about twelve years away, and he had no real idea if all the items necessary for their survival on Mars would last that long. Everything was a vital piece of their survival. No piece of equipment was discarded until every wire, diode, transistor, and rare metal was recycled.

Comms with survivors on Earth continued from time to time but were becoming rarer and rarer. Perhaps equipment was breaking, perhaps people were dying. Another rescue mission was planned, but resources were stretched so thin on Mars that it was scrapped.

As Kwame entered Mission Control, Bradley Osmond greeted him. He had been rescued from the Groom Lake control bunker. Bradley was an engineer, his specialty in guidance and propulsion. His domestic partner, Calvin Murdock, worked with him. He was an instrumentation engineer, who had worked topside at Groom Lake. Bradley and Calvin had led the small team on Mars that triple-checked the ejected-steam braking concept. Their input to the effort went further than checking out the concept; they also checked into the implementation. They believed the steam ejection would be inherently variable. For this reason, shutting off and opening the valve to the nozzle in a precise manner was a requirement to keep the *Calypso* properly oriented during the steam release. Otherwise, the line of force of the ions exiting the main nozzle would allow the *Calypso* to drift off course, with disastrous results. In the end, it was decided to make eight smaller nozzles, spaced circumferentially around the main propulsion nozzle. This would allow for better control of the steam.

Kwame's Mission Control was in the observatory room, which housed three different telescopes. As Calvin was telling Kwame about the observations of Earth made in the previous eight-hour shift, the chief astronomer Dr. Jillian Johansen came into the room and began talking with one of her junior astronomers.

"Well, if it isn't lunar crust, what is it?" the younger colleague asked Jillian.

"It's shiny. Too shiny," Kwame somewhat sheepishly answered from across the room. "It was reflecting a ton of sunlight. Additionally, it changed orbit since your last observation at midnight. I couldn't sleep. I was here last night from midnight to three. I hope you don't mind, but I wanted to see it myself."

Jillian answered back, "No worries, Kwame."

Just then, Kwame got a comm message on his wrist communicator. "It's the comm room," Kwame said as he looked at Jillian. "The base director, Gus Johnson, is asking you and me to report there ASAP."

When they arrived at the comm room, Gus was waiting there for them with several others.

"We've received a message," Gus announced, "from the moon."

"Lunar base one or two?" Kwame asked, confused by the fact that there would be any message from the moon bases.

Gus answered, "It came from one of their sailing ships that was at O'Rourke when all hell broke loose, the FBS *Liz Cheney*."

"What did they say?" Kwame and Jillian asked in unison.

"They tracked a ship that left the moon," Gus said. "Just prior to the events that destroyed everything."

Kwame asked, "Did the *Liz Cheney* know where it was headed?"

"Not exactly. They lost track at about one year when the ship seemed to disappear into an area of dark matter," Gus answered.

Kwame's eyebrows were furled as he stared at the floor; then he said, "A wrinkle in space-time."

"A what?" Jillian asked, perplexed.

"The *Calypso* has discovered, without going into the long story behind it, a map, if you will, that shows where 'wrinkles' are located. They hide starlight from stars behind the wrinkle. They hide a ton of stellar mass. We have called the unseen mass 'dark matter,'" Kwame said. "The wrinkles are simply rebounding movements from the stretching of space-time as it expanded after the Big Bang or the initial time and place event," Kwame said. Then he asked, "Could we here on Mars follow their ion exhaust track somehow?"

"No, we couldn't from here on Mars," Gus answered then paused momentarily. "We would need a ship to go out and detect then follow the ion trail. So the *Liz Cheney* tracked this ship into interstellar space? Do we have ships capable of a journey into interstellar space, I mean, besides the *Calypso*?"

Kwame frowned then said, "No, just the *Calypso*, as far as I know."

Jillian was deep in thought, staring at the wall. "We could use the *Centauri Observer*," Jillian said almost as if she was thinking rather than speaking.

Kwame straightened up in his seat when he heard this. "What is the *Centauri Observer*?"

"It was a Mars Science Foundation craft built for a voyage to Alpha Centauri. It was completed by Baltimore Aerospace on Earth and brought here for software installation and other work. It was never launched because of the events that destroyed the Earth," Jillian explained.

Kwame thought for a moment then asked, "Was it designed to be a manned craft?"

"It has three hibernation tubes, but we did not want to make it a manned mission," Jillian explained.

"But we could, right?" Kwame asked.

"Yes, but it would be a highly risky mission. This group of aliens is clearly hostile. What do they do when, or if, they catch up with them?" she asked.

Kwame answered, "We would not engage them. We want to gain intelligence on them. We want to know where their planet is located. We want to know their capabilities, etc.," Kwame added. "Let's take a deep dive into this possibility, but we should try to do this quickly. If they get too far from us, we may not be able to reacquire their trail. Jillian, can you get a team together and investigate a manned mission using this spacecraft?"

"Sure," she answered.

"Good, let's say two weeks, then we can get back together and discuss this further," he said.

CHAPTER 69

The Mother Speaks

TERRA WENT TO her room. She sat in a chair and began singing. Within a minute or two, The Mother appeared in her room.

"Mother, we are going to use most of the water in our storage to gain additional braking thrust. It is normally needed for personal use and for cooling our facility reactor. Without it, we can't generate power to maintain the hibernation tubes and other vital activities. If we establish orbit around your world, will you be able to transport water up to our ship to replenish what we used for braking?" Terra asked.

"We have space elevators to lift very heavy payloads into low to mid-orbital ranges. At worst, we can lift water to our shuttle craft and deliver what you need. At best, you can position your ship in a midorbital range, and we can deliver it directly. I want to let you know that you will need to remain on our world for up to six of your weeks. We need to train you in your duties for assisting in the development of the native peoples on your new world, and we will need to stabilize your cellular biology. You will need to pick a small group of people who will train with you also.

"Also, the AOW attacked our world and believed they destroyed it. We have moved our planet to this new orbital plane and camouflaged it to prevent further attacks. Because of these measures, it is difficult to establish an orbit around our world. We will send a craft to meet you once you are closer and assist you in the orbital maneuvers that you are attempting. Terra, I am looking forward to our time together," The Mother finished.

CHAPTER 70

Hot Pursuit

ALL THREE WERE volunteers. All three knew that the chances of returning to Mars were, well, unknown. The "plan" did not specify how or when to return to friendly rocks. Their intention was to observe, gain intelligence, and report it back to Mars then find a way to return. They would be spying on a race of people with significantly superior technology. Only they could decide when they had enough intelligence and could return home.

Tariq and Shandra were huddled over a monitor that displayed xenon isotopes, which they hoped were from an ion propulsion engine. They hoped it was from the craft that left the moon with the AOW personnel shortly before the Earth was destroyed. It had only been a few days since they came out of hibernation, but already, tempers were flaring.

Carl Peterson felt left out standing behind them. He asked, "Is this where their ship entered the wrinkle?"

Shandra turned toward him and answered, "We think it is."

Carl was clearly worried. "So what should we do? Go into this wrinkle, heedless of the obvious danger to us?"

Tariq turned and answered sarcastically, "Well, Carl, that is kind of the nature of our mission—it is inherently dangerous, you know, bro? Why don't you go play some video games, Carl? Get your mind off all this stealthy dangerous shit."

"OK, I knew you guys didn't want to have me on this little covert mission," Carl said, "but I am the only person alive who knows some of these people we are chasing."

"Yes, I know your motives were pure and you are just trying to save the species, Carl," Tariq added, seething with anger. "But you must remember that you played a part in causing all this!"

"Tariq! We're going to be together on this ship for a long time. We must get along. Carl didn't want this to happen any more than we or anyone else did. We are here now, in a tough spot, together, for better or worse. Let's play nice!" Shandra said, exasperated.

"OK, Shandra. You know I am still processing my grief," Tariq said, his voice quivering even as he raised his volume. "My entire family was killed by reckless actions set in motion by this cracker!"

"Tariq, I know you hate me and you blame me, but that is just not fair. I was just trying to obtain alien technology for our species, I—"

"You were trying to make your second trillion dollars because one trillion wasn't enough!" Tariq said as he turned and faced Carl, menacing him with his imposing physical size.

"OK, I'll leave you two to your science experiment. I'm going to go lie down. Maybe I can dream about when the world was still here, and I could buy anything or anybody I wanted. I don't want to dispossess Tariq regarding my motivations!" Carl said as he descended the ladder to the living quarters inside the ship.

It had been nine months since they began their pursuit of the fleeing AOW ship. They had spent most of that time in hibernation until they reached the location where the AOW ship seemed to have disappeared into a wrinkle. A faint ion trail was now all that they had to use in following the ship.

"The ion trail ends soon. I suppose we continue our course on this line of ions and enter the wrinkle?" Shandra asked.

"Seems like the only thing we can do," Tariq said softly, though he was deep in thought, having calmed himself down by now. "I guess we plunge through it then, see if we can pick it up on the other side."

"OK, let's do it," Shandra said.

She had been put in command so she knew she had to voice her opinion, but in this case, opinions were all they had.

CHAPTER 71

The Cervantes Star System

THE *CALYPSO* HAD technically reached the heliosphere of the Mu Arae star when the deployed solar sail failed. Six months later, the star was easily visible to the naked eye in the observatory. Crew members frequented the observatory, hoping to catch a glimpse of the star around which their new home orbited. The gas giants in the system could be seen through telescopes, but the rocky Earth-sized planets all orbited in an atypical plane sixty degrees off the plane in which the gas giants orbited. They were too small to be seen from this distance.

The *Calypso* would approach the first gas giant while applying increased thrust from the ion engines and a pair of huge solid rocket motors to slow themselves down. The gravity from the gas giant would pull the ship toward it. As the ship came to its nearest approach, the friction from the atmosphere would slow the ship even more. When the ship began moving away from the gas giant, another set of solid rocket motors would provide thrust to move the ship out of the ecliptic plane of the gas giant and into the ecliptic plane of the rocky worlds. The *Calypso* would vent their steam to slow down even more and establish a long elliptical orbit around the star and the outermost rocky planet, Terra Nova, as it was now called. At this point, a ship from The Mother's world would rendezvous with the *Calypso*, and The Mother's people would navigate the *Calypso* into an orbit around The Mother's world, which would be their first destination.

Light speed comms became somewhat practical at this proximity to The Mother's world, and comms had, in fact, been established. It turned out that the controllers on The Mother's world were fluent in English. They had, after all, been listening to broadcasts from Earth for centuries now. The Mother broadcast a message that was shared with the crew. In it, she introduced herself, her father and mother, Elotim and Mareye. She spoke from inside her living quarters and tried to anticipate what questions the crew might have for her. The distance to her world was still far enough that the signal travel times made a normal question-and-answer session impractical. That would change over the next three months as they approached closer.

The Mother's world was 0.85 the mass of the Earth and approximately the same density. This made gravity conditions slightly advantageous for the travelers from Earth. The atmosphere was slightly richer in oxygen which would also be beneficial for the Earthlings. It turns out that the evolution of intelligent life on planets required size limitations on the planets where life could evolve. The reasons were not necessarily obvious. The capability to develop space travel required that the planet have a gravity that would not be too large as it would be impossible to develop materials and propulsion techniques that could lift payloads off the surface. The intelligent life pattern was always *Homo sapiens*–like, though differences could always be found among the life that The Mothers created in the universe. Having hands that were dexterous enough to mine minerals and metals and perform intricate fabrication of scientific instruments like telescopes, electronic devices, and such required human-like hands. These were just a couple of the reasons why life was chosen to begin only on planets that bore a similarity to the Earth.

The day finally arrived to put the steam-venting concept into use. Captain Simmons, Peter, Stuart, Gunner, and Sam were on the bridge as the countdown began for the steam to be released. Lights in most parts of the ship were off since they only had enough water for the propulsion reactor, which produced gaseous xenon ions for the ion propulsion engine. The emergency batteries and solar panels topside were the only sources of power for the ship's various systems, including life-support functions like removal of CO_2 from the ship's atmosphere. Once they vented the steam, the clock would be ticking until the batteries died. They would have about three weeks of emergency power; it would take two weeks to arrive at The Mother's planet, where water was waiting for them in a storage chamber atop their space elevator. The lives of everyone on board and the thousands still in hibernation depended on the quick transfer of coolant water and the restart of the facility nuclear reactor.

"Three, two, one, venting!" Tension filled Stephanie's voice as she counted down over the intercom. The ship began shuddering slightly, and everyone felt a stronger pull toward the floor as the ship's deceleration increased from the extra thrust from the steam. It was going to take about thirty minutes to vent the steam. On the bridge, Captain Simmons and the others looked at the big screen, which showed a continuously updating plot of ship velocity. A horizontal line drawn across the plot showed the velocity they needed to achieve to make an elliptical orbit around the star and outermost rocky planet. At first, it seemed like the velocity hardly dropped at all; but within five minutes, the velocity began dropping steadily. By twenty-five minutes, the velocity had dropped to the desired level, and they stopped ejecting steam, with some still left in the boiler.

"Stephanie, we terminated early by five minutes. How much water do we have left?" Peter asked over the comm.

"Looks like about 1.5 million gallons. We should get a reactor engineer to calculate how long we could run the facility reactor on that amount of coolant water."

“Roger that, Stephanie. Good work, everyone. Peter, can you get that estimate going with the reactor engineers?” Captain Simmons asked.

“Will do, captain.”

CHAPTER 72

Rendezvous with The Mother's Port Pilots

IN THE COMMUNICATIONS room, Sam was communicating with The Mother's ship.

"*Atlantis*, how do you read us?" Sam asked over the light speed comms.

"Acceptable, Calypso. We have you on radar. Docking bay is on your port side, correct?" came the reply from The Mother's port pilot ship, the *Atlantis*.

"Correct, *Atlantis*," Sam replied. Thirty minutes later, the *Atlantis* was inside the giant shuttle bay of the *Calypso*.

Sam handed off his communication duties to his colleague and headed down to the hangar bay to be part of the greeting party.

Captain Simmons, Timeye, Sam, Riley, Terra, Peter, and Stuart stood together in a line in the giant shuttle hangar bay, awaiting the opening of the door of the ship from The Mother's world.

Peter whispered softly to Stuart while they waited for the door to open. "The name of their ship is *Atlantis*. Does that mean,. perhaps, that there was, in fact, an actual place called Atlantis?"

"It seems like it does," Stuart answered softly.

The side door of the craft opened with a slight hiss caused by a pressure differential between the *Atlantis* and the shuttle bay. Two men and four women walked down the ramp in a single file.

The men were tall, about 200 centimeters. The four women were closer to 177 centimeters, also tall. Captain Simmons was 173 centimeters, a tall woman in her own right, but clearly shorter than these women. All six seemed to be about thirty-five to forty years old. The men wore a very dark-blue single-piece jumpsuit; the women wore dark red jumpsuits. Over the left chest of each person was a name, written in the Latin or Roman alphabet universally used on Earth in the twenty-fourth century.

One of the men stepped toward Captain Simmons and extended a hand.

"I am Cleo, the pilot of this craft and an envoy of The Mothers. I understand you are Captain Sandra Simmons, commanding officer of the *Calypso*. Greetings, welcome to the Cervantes star system."

"I am pleased to meet you. Welcome to the *Calypso*," Captain Simmons said as she extended a hand to shake his. Each group then introduced themselves and shook hands with each other.

"I understand you have about two of your weeks of power left in your energy production systems on board. Entry into orbit around The Mother's world is a complicated process—it will take all of that time.

We have on board an energy device that will provide you with power for six of your months. There will be no need to hurry this process. The time-consuming nature of establishing orbit is due to the camouflaging and defensive systems around our world that we have constructed. As you know, the enemy of The Ones, the AOW, has already attacked our world though it was more than twelve thousand of your years ago. *The attack stunned us so much that we moved the location of our world.* We will work with your personnel to ensure that your ship is safely brought within the defensive system. We have two personnel who will help tie in the energy device I spoke of so that you will have the power you need to preserve the corporeal life-forms that you have in hibernation and the other energy needs you have as well, such as life support. We are prepared to start tomorrow on these tasks. We are weary, as this hour is normally in the middle of our night. We would like to rest for several hours. May we be taken to the quarters you have arranged for us?"

"Yes, of course. Terra, Sam, and Riley, could you take our guests to their quarters?" Captain Simmons asked when she turned to Terra and her parents.

"Of course," Riley answered.

"Cleo, eight hours from now is our normal evening meal. Would you be willing to join us for that meal in our assembly room? We were given information regarding your dietary needs, so we hope we won't cause you any digestion issues right off the bat," Captain Simmons said.

"Yes, captain, we would be honored to join you then after a rest," Cleo said as they were led off to their quarters.

CHAPTER 73

Dinner with the Creators

THE MAIN ASSEMBLY area was also a cafeteria, which was opened on demand. Meals could be eaten here or in personal quarters. A small group of cooks prepared meals for everyone on board. This ensured that the optimum dietary requirements of everyone were maintained. In the case of this meal, the visitors ate entirely fruits and vegetables. The same group that greeted The Mother's people would be sitting at two large tables with them.

The visitors were staying in a vacant apartment, and they came as a group to dinner.

Cleo sat down with the captain, Timeye, Sam, Riley, and Terra. Two of the female visitors also sat down with Cleo and the *Calypso* crew members. The others sat with Peter and Stuart at the other table. Cleo introduced the two women Rachel and Kristine again and said that they were his soul partners.

"What is a soul partner? Is it like what we call a spouse, a husband or wife?" Terra asked.

"Not exactly," Rachel answered. "Cleo, Kristine, and I were split from the same portion of life energy at our beginning. We lived many lives together before we were born in The Mother's world. Sometimes we were married, sometimes we were friends or family members together."

The *Calypso* crew members all looked at one another with confusion.

"I sense that you are in wonderment at this answer. Many of your mystic traditions never postulated that we live many lives nor accurately described what we call the Wheel of Progression through Lives. Each of you has lived many lives before this life that you are currently in. Sometimes, you have been with your soul partners as a spouse. We are not allowed to tell you much more information than this. It would interfere with your progression in this life. Terra, however, is a blue-level soul, an advanced soul. She will be given all this knowledge when she arrives in our world. Each of you will progress to that level eventually, not to worry. So much of what you 'know' concerning these things was wrong. All of you will be taught about the Wheel. It will bring you peace and comfort that you have lacked," Kristine explained.

"When learning things like this, it is best to learn small amounts at a time," Cleo explained as he began eating an orange. The others began eating as well though it was easy to sense that there were a great many questions in the minds of the crew at the table.

CHAPTER 74

Capture

THE *CENTAURI OBSERVER* had reacquired the ion trail after they passed through the wrinkle. There was no way to tell yet where they were; all they knew was that the star field looked different. Tariq was working on determining their location. He was not confident that he would figure it out.

"I just thought of something, Shandra. Kwame mentioned that the *Calypso* had acquired a map device thingy. Maybe I could describe the star field I see and ask Kwame to see if the *Calypso* can figure out where we are," Tariq said.

Shandra replied, "Sounds like an idea worth trying."

"So we're lost!" Carl shouted. "Jesus H. Christ! We'll never get home!"

"Carl, you need to remember we don't have a home anymore, thanks to you," Tariq said harshly.

"Guys! Leave it be," Shandra said with exasperation. "It's not helping anything for you two to keep on like this!"

"OK, I've described the star field with coordinates relative to our ship and its location. I put it in table form with brightness and locations. Let's see what they can tell us," Tariq said as he pushed Enter on the EPC device.

"Tariq, I've got something on the differential gravitometer. Come take a look," Shandra said with urgency.

"It could be an asteroid, but yes, it could be a ship also. A big one," Tariq said with concern. "I say we go to full thrust and go back the way we came, across the same wrinkle. We aren't stealthy, and they must be. I don't like this situation."

A voice came through the comm. "*Centauri Observer*, stand down. We are going to pull you into our shuttle bay. Any attempt to flee will result in your deaths!" The voice on the other end was filled with menace.

"Tariq, get a message back to Mars!" Shandra shouted. "Tell them we have been located by the AOW and we will be captured. Do it now!"

"Copy that," Tariq said as he tried to type a message into the EPC.

Just then, a tractor beam jolted the ship. A blue light penetrated the skin of their vessel, and Tariq dropped the EPC device. All three crew members fell to the floor unconscious. The *Centauri Observer's* electrical systems failed, and their craft began to be pulled into a cavernous bay that opened on the side of the giant AOW ship.

As the tractor beam shut off, the *Centauri Observe*r dropped a meter onto the deck, jarring the occupants. All three crew members regained consciousness.

Everything was silent for a moment until the same voice came over the radio. "Do not touch your EPC device. Any attempt to do so will result in your immediate termination. Exit your craft immediately."

"Is the atmosphere suitable for us to breathe?" Shandra asked, speaking into the radio.

"Yes," came the reply.

Tariq and the other two stood up and walked to the airlock. They entered the air lock and then opened the outer door. As they exited their ship to the deck of the shuttle bay, they saw a single light on the wall in front of them. The light then began moving toward them. The ball of light, about the size of a basketball, first came over to Shandra and shone a blue beam into her eyes. Shandra fell to the floor, unconscious. The ball of light moved to Tariq and repeated the same sequence of events. Tariq fell to the floor unconscious. Carl shrieked in terror and began to run back into the ship, but he was stopped by the same blue light and fell off the ramp to the ship, causing him to break his right arm just before he went unconscious.

CHAPTER 75

Interrogation

SHANDRA SLOWLY WOKE up. She was dizzy and in pain. She got to her knees and then stood up slowly. She was in a room without any light, except for a very small blue ball of light that hovered in the corner of the room. She looked around the dark room and did not see Tariq or Carl. She suddenly realized she was naked and cold. Apparently, her clothes had been removed while she was unconscious. Her head throbbed, and her neck was so sore she could barely turn her head. As she put her hand to her left temple, which was where she noticed the most pain, she realized there was a small device implanted in her head under the left temple.

Shandra sat back down on the floor, wrapping her arms around her breasts, trying to stay warm. After a few minutes, she became frustrated and shouted out, "I'm cold, I'm in pain! Can I see a doctor, please!" A moment passed, then a door opened on a wall and a man walked in.

"Who are you?" she asked.

The man replied, "I am a doctor. I can help you, but you must help me first."

"What do you need from me?" Shandra asked between tearful sobs caused by the pain in her head.

"We need to know where the *Calypso* is headed and how many blue souls they have on board," the doctor replied.

"I don't know the destination of the *Calypso*. It's apparently a secret. I don't even know where it's presently located, and I don't know what a 'blue soul' is. What's a blue soul?"

"If you don't know what a blue soul is, then you can't help us. If you can't help us, you are of no use to us, and we will terminate you," the doctor answered.

"Wait, wait, is a blue soul a mind reader?" Shandra asked.

"Yes, but they can do so much more than that. How many blue souls are there on the *Calypso*?"

"I think just one. I've heard that there is a girl who is special, but that's all I know," Shandra answered. The pain in her head seemed to increase tremendously in an instant. "Please, can you help the pain in my head? I can't take it, please help me!"

The doctor took something out of his pocket and pressed a button. Instantly, the pain in Shandra's head disappeared.

"Thank you, thank you," Shandra sobbed, relieved of the unbearable pain for at least the moment.

"We want you to work for us. If you do this, you will live, and the pain will never return. Will you help us?" the doctor asked.

"Yes, yes, I'll do whatever you want," Shandra answered.

"Good. We are going to have you return to Mars and work for us. Your cranial control device will give you instructions via a miniaturized EPC device. You will get the information we ask for. You do this, and you will live. All your people on Mars will live, and you will join the AOW. We will help you restore the Earth, but you will serve us. You will help us fight The Mothers," the doctor explained.

"Who are The Mothers?" Shandra asked.

"They are a race of beings who enslave humanity all over the galaxy and beyond. Helping us defeat them will set you free from their oppression," the doctor said.

"But didn't the AOW destroy the Earth?" Shandra asked.

"Absolutely not. This is misinformation that The Mothers put out to cover up whenever they are finished with a world and choose to destroy it," the doctor said.

Shandra was led down a hallway to a similar room where Tariq was waiting in the same condition. "Where is Carl?" Shandra asked the doctor.

"He has been terminated. Those who work for us and then turn against us are tortured and killed, very painfully, very slowly," he said.

Tariq and Shandra looked at each other with shock.

"Here are your clothes." Another individual entered the room and threw their clothes on the ground in front of them.

"You will get back on your ship and return to Mars. You will tell them that Mr. Petersen was killed when he tried to commit suicide by depressurizing your ship. You survived the event, but he did not. You will tell them you lost the trail of the AOW ship you were tracking and returned without ever meeting us," the doctor said.

CHAPTER 76

The Entry Protocol

THE *CALYPSO* WAS attempting to set up an orbit within the defensive measures around The Mother's world. Captain Simmons had given Cleo's team full control of the ship in terms of navigation. The ship Cleo's team had arrived on had a supercooled gas that was released outside of the *Calypso*. Cleo explained this would make the ship look like a comet if the AOW was looking for any ships in this area of space. The AOW was actively searching for exoplanets that could contain life started by The Mothers. This technique and others were used to disguise spacecraft arriving and leaving The Mother's world. This was done because the AOW and other civilizations such as the Earth, for different reasons than the AOW, were constantly looking for exoplanets. The Mothers wanted each world to develop independently of other worlds they created.

Peter, Stuart, and Stephanie sat with Cleo on the bridge.

"Since the time of the attack on our planet by the AOW, we have moved our planet out of the plane of the orbits of the gas giants around our star. This has essentially hidden our world from detection by 99 percent of the stars in our galaxy. We have also deployed blackout screens beyond the orbit of our world that make our planet look like a gas giant to the 1 percent of stars that still have a view of the plane of our new orbit. By doing this, our world looks much, much larger than it is to anyone looking for a small, habitable rocky world. Anyone looking for small rocky worlds would see what looks like a gas giant when our world transits across the face of our star. Such a planet could not host complex life-forms," Cleo explained to Stuart and Peter.

"How do you move the orbital plane of a planet?" Stuart asked Cleo.

"Large asteroids. We used five large asteroids to pull on our planet. Over more than a thousand years we successfully moved our planet. Of course, this meant that we had to live underground in sealed chambers as the climate and atmosphere were affected during the time the planet was moved. The move destroyed all the higher life-forms in our world. At this time, we are in the process of reestablishing the original life-forms on our planet," Cleo explained.

Peter broke from deep thought and asked, "But you are safe now from detection and attack by the AOW?"

"Yes, because they think our world was destroyed," Cleo answered.

"How did the AOW attack your world?" Peter asked.

"The AOW directed a swarm of asteroids at our planet when they attacked us. We deflected all but two of them, but those two caused enormous damage. We were able to avoid any loss of human life by anticipating where the impacts would occur. Right now, the AOW thinks that we left our destroyed planet for another world. They have never actually entered our planetary system. It is too far from any of their worlds, and they lack the map of the wrinkles to get here in any reasonable time. We have established defensive outposts in the most favorable locations, such as the location where you picked up the Caldonian survivors. We have also intercepted and disabled a number of AOW warships before they got anywhere close to our system."

"This war between the AOW and The Mothers has been going on for centuries. What do they want?" Peter asked.

"Power and control. This is what results when only men rule. This was the problem on your world for so long. Your leaders were almost always men. Women were excluded by custom and/or law. This resulted in an imbalance between the emotions of anger, aggression, and revenge versus compromise, cooperation, compassion, and forgiveness. Even when this began to change in the last three centuries on your world, the tradition had been solidified that a leader must be strong and almost never compromise. So naturally, the women who got into leadership positions were women who weren't very different than the men they replaced, and the same results occurred. The result was that your world was essentially destroyed by greedy, power-hungry men who ignored the existential threat of human-induced short-term climate change. Your world has been doomed for well over five hundred years," Cleo explained.

"Wait, The Mothers are obviously all women. So in a perfect world, are only women supposed to lead?" Peter asked somewhat incredulously.

"Perhaps I have poorly explained our sociological structure," Cleo said. "Women are in charge, not always, but quite often. This is deliberately done so that women can use their advantages to achieve the best outcome for society. Men are utilized for what they do best—execution in creating a society. Men are usually, but not always, put in charge of carrying out the creation of civilization and worlds, but they work for The Mothers. These are generalizations I am giving you. There are exceptions, but I am attempting to explain why our society is structured the way it is."

"So we just did it all wrong?" Stuart asked.

"The AOW turned a number of men who oversaw the shaping of your world. They allowed the evolutionary progression of male dominance to continue as your species was moved from carnal to virtuous life-forms. This doomed your society to endless cycles of wars, famines, stunted technological development, and so many other problems. However, none of these compares to the loss of divine sight that resulted from a rejection of The Mother's teachings and then an outright ban on their teachings. I realize this is a lot to process but let me continue to try.

“The mystic teachings that were developed on your world were also male dominated. You created gods and almost always had a dominant male god. The evolution of this type of male-dominated system of mystic belief became especially pronounced with Christianity. In that belief system, there was only one god, a man. Your Christian scriptures have very few female references in terms of characters, and they give almost no significance, essentially, to the roles that women played. All of this tended to significantly hinder the confidence, opportunities, and expectations of women, to the detriment of the survival of your species. One could say that it is like choosing to deliberately use only one-half of the brain power of your species,” Cleo finished while Peter and Stuart sat quietly, deep in thought. Stephanie, however, was smiling broadly.

CHAPTER 77

Arrival

AFTER TWO WEEKS of maneuvering, Cleo and his team had the *Calypso* in a high orbit around planet number one, The Mother's world. Since this was not within reach of any of the space elevators, the delivery of water stores was going to take another two weeks as the shuttles would be busy ferrying water from the elevator to the ship.

The Mother asked for a small team to come down to her world. It was decided that Peter would remain behind to oversee the ship operations and water delivery. Cleo and his team would remain on board to guide the *Calypso* to the settlement planet though that would only happen once the landing party was finished with whatever The Mother was going to do with them. The landing party consisted of the captain, Terra, Timeye, Stuart, Sam, Riley, Hatcher, Kathy, and Kaylee. They would descend to the surface in one of the *Calypso*'s shuttles. Hatcher and Stuart would pilot.

The Mother's world looked so very much like the Earth. There were large oceans and clouds. Complex landforms like mountain ranges and large fertile valleys and river systems were observable from orbit. Many cities could be seen though the lighting was different from the cities on Earth. The cities seemed to glow, even in daylight.

The descent to the capital city was controlled by an explicit set of instructions that were fed into the shuttle's navigation computer. As the shuttle left the hangar bay and moved away from the *Calypso*, they realized that this was the first time most of them had set foot on a world in years. They all viewed the *Calypso* as a ship and not a world such as Earth. The excitement was palpable. The navigation computer took over just after the shuttle left the hangar bay, leaving Stuart and Hatcher with very little to do. They would only be necessary if anything went wrong, and Cleo assured them nothing would go wrong. Apparently, nothing had gone wrong for a very, very long time.

"Terra, did The Mother say exactly who would be there to greet us?" the captain asked.

"She didn't say anything. I assume herself, of course," Terra answered.

"I wonder who among us will be given the initiation rites?" Timeye mused.

"Timeye, what are the initiation rites? I think you presume we all understand more about The Mothers than we actually do," Riley said.

"I assumed Terra knew what this was and had explained it to you," Timeye answered. "At this point, I think it is best to let The Mother explain it to you."

"We are going to enter the atmosphere in a moment. Everyone, be sure you are secure in your harnesses. Don't forget the head restraint. If we have a hard landing, don't worry, your seats have ten-g energy dissipation in their mounting points," Stuart explained.

"Thank you for that tidbit of information, Stu. I wasn't worried at all until you saw fit to remind us that this is always the most difficult and dangerous part of space travel," Sam said dryly.

"Relax, everyone. The probability of a successful landing is almost acceptable per Caldonian requirements," Timeye explained.

Suddenly, Stuart projectile vomited all over his side of the control panel.

"Eww! Can the ventilation system be boosted to help with the smell?" Terra asked, looking directly at the clearly embarrassed Stuart, with a half-smile on her face.

"I'm turning up the fan speed and introducing a fragrance to the air handling system," Hatcher said.

"Thank you, honey," Terra said, looking over at Hatcher.

The ship began to shake as it encountered the upper reaches of the atmosphere.

"Braking chute deployed. Six minutes to spinner engagement," Hatcher said, taking over for a still-recovering Stuart. A few minutes later, a groaning sound could be heard and felt as the spinner slowly began rotating.

A few seconds later, Hatcher announced, "Braking chute cut, spinner at 90 percent. Things should be less exciting now." As he said that, the bumping and swaying dropped significantly as the shuttle was now in powered flight.

"Altitude seventy-six hundred meters. Descent rate nine hundred ten meters per minute. About eight minutes thirty seconds to touchdown. Sit back, relax, and enjoy the view, everyone. We thank you for flying with *Calypso* Shuttle Services, and we hope that when your plans call for planetary descent or ascent, you choose to fly with us again," Hatcher said as he turned and faced everyone with a broad smile on his face.

The descent continued for several minutes then slowed down considerably. A clear view of the ground appeared as they passed through the lowest cloud layer. Below, a large courtyard could be seen with grass at the center. A city had been built around this courtyard. It was daylight now where they were landing. Even so, the ground looked oddly illuminated, as if a source of light were in the ground itself.

The shuttle glided silently to one end of the large grassy field and slowly descended. "Sixty meters, descending. Touchdown. Spinner shutting down. Folks, let's wait for the all-clear on bioassessment," Stuart said, having regained his composure in the last few minutes of descent.

A couple of moments went by while the atmospheric composition and bio content, temperature, and radiation levels were measured. The ground-penetrating radar showed solid ground under the shuttle with approximate ambient temperature, making it equal to the air temperature inside the shuttle.

"All right people, a couple of things," Captain Simmons said. "Terra will exit first, followed by the rest of us. Terra will take charge of our greeting protocol. We'll follow her lead."

"Whew, I am nervous," Terra said quietly to her mother as she stood.

"I know you are, honey, but you'll do just fine," Riley said reassuringly to her daughter.

They descended to the lower level of the shuttle where the airlock and ramp were located. Terra stood by the air lock control and hesitated then punched in the ENTER code. The inner door slid open, and they all stepped into the air lock. Terra entered a manual open code to the outer door, and it slid open. The sunlight seemed very bright, almost too bright, as each of them raised their hands to shield their eyes. A moment passed before their eyes adapted to sunlight that they had not seen in years. Terra turned to look at her parents then started down the ramp with everyone else following silently in a single-file line.

When they reached the bottom of the ramp, a surprise awaited them as the section of grass they stood on lifted upward about half a meter and started moving them quietly toward the other end of the long courtyard. As they reached the end of the courtyard, the flying section of grass they had been riding settled down and seemed to be absorbed into the grass below it. A building that ran the length of this side of the courtyard stood before them.

A door opened, and a woman walked out and over to Terra. She spoke in a clear and feminine voice. "I am Eve of Eden. I welcome you to Preemus. Terra, I have long awaited and anticipated your *return,* beloved and dear *daughter.*"

CHAPTER 78

Trying to Drink from a Firehose

RETURN AND *DAUGHTER.* *What did she mean by those two words?* Terra wondered with an obviously puzzled look written on her face.

I will explain that later when we are alone, my daughter. Those words seemed to pop into her head as The Mother stared into her eyes. The Mother then began greeting each other by name and then with a hug that put all of them at ease.

"Come, let me show you our capital city. Don't worry, I will show you just a little of it. I know you are all weary. You have been inside a spacecraft for so many years, too many new things all at once can be a bit tiring," The Mother said.

"How should we address you?" Captain Simmons asked.

"Just Eve," The Mother replied.

"We would love to see the city, especially if we get another magic carpet ride!" Hatcher said playfully.

"Yes, Hatcher, we will ride on another magic carpet." Eve smiled.

Just then, all of them began to rise in the air, catching them by surprise again. "Don't worry, Hatcher, there is a 'magic carpet' under us. It just can't be seen like the grass, but it is there," Eve said.

They rose to a height where they could see the entirety of the city. There were buildings large and small, but they were different in very odd ways. It seemed as if the buildings were alive and growing even though none of them could see the buildings getting larger.

"Our structures are made of living organisms. We ask them to grow in certain ways so that we end up with a structure for our housing. In turn, we provide them with what they need to grow and fulfill their purposes. We live in symbiosis with the other families of life," Eve explained.

"How do you communicate your needs to a plant or tree?" Sam asked.

"We feel each other's life energy," Eve explained.

"What is 'life energy'?" Stuart asked.

"Since you are a physicist, you are familiar with what you used to call the Higgs field. This gave mass to subatomic particles that you used to call bosons. There is a boundary, somewhat like this field, between the energetic universe and the physical universe. Most of all the life energy in both universes resides in the universe of energy. A small portion of that life energy resides with our physical forms here in the physical universe. That small portion of energy within each of us is our connection to the universe of energy. Each universe is a mirror image of the other universe. There cannot be a physical universe without an energetic universe that was ordered first. The physical or mass universe grows out of the energetic universe. This field is the boundary layer between universes. Everyone born here has the ability to converse across this boundary to the energetic universe. You have called this 'mind reading' at times, though that is a gross understatement of what this ability really is. The universe, galaxies, and worlds were organized using this ability," Eve explained.

At this point, the conversation stopped as each one of them contemplated what they had just heard.

"The only thing we couldn't do was to set the initial conditions of the universe at the initial time and place. To establish the universe at the lowest point of entropy at the initial time and place required input from The Ones, those who rule in the realm of energy, the energetic universe," Eve explained.

This was followed by more silence. Finally, Riley said, "Terra can read minds, but she wasn't born here."

"Yes, I will explain more on this later. Let us now go to your quarters. I know it is late in the day for each of you. You can rest for a few hours, then we will meet for dinner," Eve said.

"I feel like someone just turned on the firehose of knowledge and answers. Do you drink alcohol here? I think I might need a drink," Stuart said as they all broke from their stunned silence and laughed.

As dinner with Eve, Terra's parents, and her younger twin sisters, Kathy and Kaylee, and other *Calypso* crew members ended, the *Calypso* crew started filing out to return to their quarters.

Terra thought to herself, *The length of a day on Preemus is twenty-six Earth hours. Our contingent from the* Calypso *is on a completely different work/rest schedule compared to here. I don't think I can sleep right now.*

"Terra and Hatcher, I know neither of you are weary. I'd like to talk with both of you. Can you come with me?" Eve asked.

"Yes," Terra responded as she looked over at Hatcher.

Eve led them down a long hallway with a stone-cobbled floor. They entered an empty outside courtyard that was bounded by a forest at the far end, more than one hundred meters distant. All three stopped when they reached a spot with what looked like a firepit.

Eve sat with her legs folded to one side and an arm propping her up.

"Please come sit. We can talk now in private."

Terra and Hatcher exchanged a questioning glance then sat down.

As they sat, Eve began "speaking" though no sound came from her mouth. *Both of you will perform important work in the development of a new world, which will be called "Terra Nova." I speak of the second planet within the new ecliptic plane of this star, Mu Arae or Cervantes. This will not be the final destination for your people.*

Instead, they will perform the necessary work to develop an intelligent native species on Terra Nova. This species and this world will be given to new souls. Our purpose is twofold: first, to complete the species development, and second, to help you learn how to do this. The Calypso *will be emptied except for a crew necessary to get it back to rescue your people on Mars. Upon completion of this work on Terra Nova, thousands of your years hence, your people will return and liberate the Earth from the AOW and restore it to its original state, as it was in the beginning. Your people will reside again where they originated from.*

Terra, you have previously lived on Preemus. You were my daughter. You had not been given the ability to remain in the physical universe indefinitely so you lived, aged, and died here. Your rebirth on Earth to Sam and Riley allows you to complete your final incarnation, which includes this very important work. You will never age beyond your thirty-fifth birthday.

Hatcher, you will live to a great age and then pass away. You will be reborn in the new world with new work and more progress to be completed. You and Terra and another will be together upon completion of your incarnations. You are two of three souls split off from The Ones. The third soul you have not yet met. Terra, I know you have many questions, but more answers must wait. Communication with Preemus will be only through you (Terra) from this point onward. You are the leader of your people, but you must give them the choice to assist in this work. Use their knowledge and our help to complete this task.

Eve finished as she looked at a stunned Terra and Hatcher.

I sense you are still unsure. You will leave for the new world as soon as you have reprovisioned your vessel with water and you have completed the training on species development. Once everything and everyone has been transported to the surface of Terra Nova, a small crew will remain on the Calypso and return to rescue survivors in your solar system. Terra, at a future time, when you have overcome the tests that will try you on Terra Nova, you will return here again. Hatcher, you, and the third soul split from The Ones with you, along with many of your people will receive the initiation rites and will be presented to The Ones at a future time. Upon completion of the rites, your people will have the power and wisdom to return to your world on Earth where you will establish a new world of Mothers from which you will serve The Ones. Your ultimate destination is the realm of the energetic universe, but that will happen at a far distant time, Eve said as she bade them stand and then hugged them one last time.

CHAPTER 79

The Choice

TERRA SAT WITH her conundrum in her quarters on the *Calypso*, trying to think through the issue.

The issue. How can I ask my people to settle temporarily on a new world, with the knowledge that their old world is gone, and ask them to help strangers develop their world? At the very least, I need to tell them about what Eve's plans are for the people from Earth. I need to tell them we will return to Earth someday, but will they follow me in helping to develop the intelligent species on Terra Nova? she thought.

Terra began singing *the new song*, The Mother's song that Eve had taught her and Hatcher, the song that would bring Eve to their aid.

As she finished, the room grew silent, and Eve appeared.

"Eve, thank you for coming. I have a problem. If I tell my people that we are not staying on Terra Nova permanently, some, perhaps many, will not go along with this plan. I am asking them to help the Terrans develop communities so that they may form their civilizations. I am asking that they sacrifice to help strangers, even when they see themselves as having no home. I am worried that some may not see the value in helping the Terrans when we are in need also. I feel I must tell them that we will return someday to Earth to restore it as it was in the beginning, but it will be their descendants who will be returning, not themselves. What would you have me tell them so that I may get their cooperation for this long mission of helping the Terrans before we return to Earth?"

"Terra, this is the last great lesson that you and your people will need to learn and master before they will ascend to the worlds of The Mothers. We are all connected in that we were all split off from The Ones. In the realm of The Ones, before our first incarnation, we all saw and felt what others saw and felt. In this world, only beings such as you and I have reclaimed this ability. The loss of our ability to see and feel what others see and feel gave all of us the opportunity to become true individuals, to learn, choose, and experience on our own.

"When we master ourselves, our individual becomes perfected and can rejoin The Ones. We will have overcome the desire to deceive, to ignore the plight of others, and so many other negative aspects of individuality. We will be refined individuals without the undesirable attributes of individuality. Tell them that this is their last great test. As they help the Terrans, as they live and die on Terra Nova, they will have the opportunity to make the last great push to the *worlds* of The Mothers and from there to the *realm* of The Ones."

"But what if some do not want to follow this course of action? Should I force them to do so?"

"We never force an individual to do anything. Tell them that their decision to act in that manner will be respected, but upon their deaths, they will be assigned back to the primitive worlds, to continue their growth in new lives until they voluntarily reach the same decision point that they are at now," Eve answered.

"Thank you, Eve."

"Terra, let us sing the song of *our* Mothers together now."

Eve began to sing as Terra joined in:

Ancient Mother, I hear you calling me,
Ancient Mother, I hear your song,
Ancient Mother, I hear your laughter,
Ancient Mother, I taste your tears.
You are the water that I drink,
You are the air that I breathe,
You are the sun who gives me warmth,
You are the earth beneath my feet.
Open your heart, open your mind,
Feel your soul, fly your dreams.[1]

At a crew-wide meeting, the choice was presented to everyone awake and present. They could choose

1 Ancient Mother by Sacred Earth.

to assist in the development of a virtuous civilization upon Terra Nova, or they could choose not to participate in this endeavor. If they chose not to participate, they would be given a settlement in the southern continent and could remain there until they died. They would then be reborn in the primitive worlds.

In the end, everyone chose to assist The Mother in her plan.

CHAPTER 80

Terra Nova

The Second Rocky World, Terra Nova
The New Planetary Ecliptic, Mu Arae Star System

THE NEW WORLD, the second planet of the new orbital plane of the Mu Arae system, was very much like the "old world," or Earth. From their midrange orbit, clouds, water, and land could be seen. The water was a brilliant dark blue, the clouds white, and the land either brown or green and lush. Snow covered the high northern latitude of the three main landmasses. In the southern hemisphere, the remaining continent was covered with lush green vegetation. At present, mapping surveys of the planet were underway. This necessary pause in the downloading was a good idea since it might be a long time before such helpful overhead views would be available again since the *Calypso* would be leaving once everyone and everything was unloaded. The mapping surveys would inform their decision on where to create the first colony. To that end, the leadership team for colonization was meeting to finalize possible colony locations based on the data they had been gathering.

"So it is agreed that these three locations will be scouted out by primary landing parties," Dr. Hans Larsen, the lead planetary scientist said.

"Yes, correct. Thank you, Dr. Larsen," Terra said.

"OK, people, we have the landing party rosters for each of the three locations posted up at the front of the conference room. Shuttles One and Two will leave tomorrow at 06:00 hours. Shuttle Three leaves two hours later at 08:00 hours. You have one week—let's get the best data we can!" Captain Simmons said, obviously happy that they would finally be getting eyes on the new real estate.

Hatcher and Terra were the last ones to leave the conference room.

"Be careful down there, Hatcher," Terra said with a tear in her eye. "We can only infer so much from orbit. Although we know what types of animals are down there, we don't know exactly how aggressive they will be, and well, there are a thousand other things that can hurt you."

"Terra, I'll be fine. Eve said I would live a long time," Hatcher said with a chuckle. "And I'm only twenty-two, so I have a while to go, I hope."

Terra leaned in and kissed him. "Be careful. Don't get in a hurry."

Hatcher's shuttle, which he was piloting, was the *Sally Ride.* Their landing party, under the command of

Gunner Lawson, would likely touch down first, as it was scheduled to leave at 06:00 hours and their landing site on the southeast coast of the largest landmass would be reached first. As they boarded their shuttle, the landing party waved goodbye to their loved ones assembled in the hangar bay with them. The hangar bay was cleared of people, the shuttle doors were closed, and the huge hangar doors slid upward, venting residual atmosphere as they opened.

The shuttle moved slowly under the weak attitude control jets, but it eventually flew clear of the *Calypso*. Hatcher fired the main engine to slow the shuttle down and begin its drop to the surface of the planet. Once this was completed, he engaged the landing sequence program, and all that remained was to wait for the final descent when Hatcher would again take over from the computer in order to find a suitable landing location. All ten crew members were buckled into seats on the upper deck. All of them had a good view of the looming planet below them.

"*Calypso*, we are now encountering significant atmospheric drag. Shuttle is repositioned for atmospheric transit. Expect comm blackout soon," Hatcher said as plasma buildup cut out radio transmission.

"One way or another, we are going to the surface," Hatcher said to the crew as the bumping continued from the atmosphere. Three minutes later, the bumping slowed down and evened out.

"*Calypso*, *Sally Ride*, do you copy?" Hatcher said into the comm system.

"Copy that, *Sally Ride*," *Calypso* comm responded.

"Deploying braking chute. Two minutes to spinner activation, over."

"Copy that, *Sally Ride.*"

"Engaging spinner. We have clear skies. I can see the surface even now. Altitude ten thousand seven hundred meters," Hatcher said as the descent continued and the world below began to have definition. "Spinner deployment in thirty seconds. Spinner engaged. I see a large flat area with a prominent hilltop about two kilometers from the coast. I'm going to put her down somewhere there, over."

Hatcher took over the controls from the computer. He angled the *Sally Ride* nose down to get a better view of the surface below. "Engaging exterior cameras. Each of you can see the camera views on your seat screens. If you have comments, please speak up," Hatcher said as he scanned the mountain range in front of the shuttle and the descending land opposite the coastal mountain range.

"I see a stream dumping into the ocean," Carrie Jaeger, a planetary scientist, said over the comm. "Let's look for some higher ground on each side of that stream. Might be a suitable water source."

"Copy that, Carrie. Looking for some elevated ground above the river level," Hatcher responded. "How about on top of that hill on the south side of the river?"

"Looks good to me," Carrie answered.

"*Calypso*, we are on final. We've found what looks like an elevated landing zone, over," Hatcher said.

"Copy that, *Sally Ride*. Go for landing," *Calypso* comms answered.

Hatcher swung the shuttle back and forth around the hilltop they had chosen. A clump of trees on the west end of the hill continued into a large forest as the hill descended back to the plain that continued all the way back to the coastal mountain range.

"OK, looks good to me. I'm going to land," Hatcher said, sounding almost out of breath as he did so. "Landing lights on, one more glance at the cameras. I see no obstacles. Thirty meters, twenty meters, ten meters, nine meters, eight, seven, six, five, four, three, two, one, contact light indicator illuminated. *Calypso*, Terra Nova Base Camp One, the *Sally Ride* has landed," Hatcher said into the comm set, which could be heard by the crew and the *Calypso*.

"Copy that, *Sally Ride.* Congratulations, Hatcher, well done," the *Calypso* responded.

"Deploying planetary sampling system. Should take five minutes, folks," Hatcher said to the crew. "Remain buckled in, in case we need to do an emergency ascent. I'll advise when you can get up."

"Hatcher, let's get the observation drone up ASAP, "Gunner said. "We want eyes all around us."

"Observation drone ejected, I've got a camera feed going. It's on all your seat screens now. Engaging automatic sentry mode. Gunner, you have the drone now," Hatcher said.

"Copy that," Gunner responded in a flat tone. "I've got the remote handling and viewing unit up and operating."

Carrie read through the data report from the sampling system. "We've got a data report from the planetary sampling system. Honestly, this looks like the readings I'd expect on an Earth location in proximity to a salty ocean. Checks out OK."

"OK, my job in the command seat is done. The crew is yours, Gunner," Hatcher announced.

"Carrie and Dale, you are with me. Let's go below and get on full biocontainment suits. Everyone else, get your suits out, be ready," Gunner said.

Everyone except Hatcher stood up to go below to the airlock. Hatcher was required to stay in his pilot seat until Gunner gave the all-clear for exiting the ship; this was done in case an emergency takeoff was required. When all three of the initial EVA crew were suited up, they stepped into the air lock as the inner door closed. The remaining crew, minus Hatcher, stood around the air lock window as the outer door slid open.

"Oh, forgot to tell you this, everyone. You'll need your UV glasses," Gunner said to the crew left behind. "None of you have seen natural sunlight for a very long time. I don't want you to get migraines!"

"That would have been good to know before we opened the outer door!" Dale LaPlant said satirically to Gunner.

"You'll be OK, Dale," Gunner responded. "We're only going to be out for thirty minutes or less."

All three descended the ramp to the ground below, which was covered with grass that seemed to be tightly interconnected with runners that ran in every direction. When they all reached the ground, Gunner unzipped his collection bag and pulled out a bronze plate affixed to a stone.

Carrie asked, "What do you have there?"

"It's a plaque. It has a message on it."

"What's the message?" Dale asked.

Gunner answered, "It's essentially the same message that Neil Armstrong and Buzz Aldrin placed on the moon when they landed there for the first time."

"What does it say?" Carrie and Dale asked seemingly in unison.

Gunner held the plaque out and read to them:

> ***Here, men and women from the planet Earth first set foot upon Terra Nova, July 2352, Current Era, Earth Reckoning. We came in peace for all mankind.***

PART TWO

The New World

CHAPTER 81

First Contact

EVE HAD TOLD Terra that they had developed hominids on Terra Nova and that they were gathered in small hunting groups across two of the continents. Terra Nova had been moved to the new ecliptic several million Earth years ago, long before Preemus had been moved. This meant there had been significant directed evolution for a very long time.

The move became necessary as the smaller gas giant in the original ecliptic swept inward toward Mu Arae, threatening the smaller rocky worlds. The destiny of all the closer orbital bodies was to be consumed by their star as the inward-sweeping gas giant would pull them along to their fiery death. To preserve Terra Nova from this fate, it was moved long ago. At present, the consumption of the smaller gas giant by Mu Arae was in process, and it couldn't come soon enough as the tugs and pulls of its gravity would eventually cause Terra Nova and Preemus back into their original orbital plane unless the smaller gas giant was consumed.

It was the dawn of the second day on Terra Nova. Gunner; Carrie Jaeger, a planetary scientist; Dale LaPlant, a psychologist; Pierre Dupont, a biologist; and

Penny Watkins, a ship security specialist, had departed on a long-range recon with the intention of finding and contacting a local group of hominids.

A drone, now circling two kilometers ahead at two hundred meters' altitude, had located a group of individuals. Terra had told each landing party that they were to contact the native people. Directing the final evolutionary changes to the hominid brains required that they develop a relationship of trust with them.

"I have eyes on the group," Dale whispered into his shoulder-mounted comm device.

"Hold in place until we all reach your position," Gunner whispered back.

As the other four reached Dale's location, they set down the deerlike animal that they had shot earlier that morning.

"OK, four of us will approach the group with the deer carcass. We'll get within fifteen meters, set the deer down, and gesture toward it. Then we will hold out our arms to our sides with our palms open. The intention is to show them that we are friendly. We will then back away, always facing them as we do. It is possible that they will follow us back to the shuttle. We want them to do so. Our objective is to make them unafraid of our technology; to show them we will help them and give them useful things," Dale explained one last time as he had briefed them the day before on the contact plan.

"Penny, you will remain behind hidden but with eyes on us as we execute this first contact. If there are problems, your first response will be to create a diversion with the flash-bang grenades. Don't throw them at the group—throw them in another direction. We will then make a retreat to this location if possible. Deadly force is not authorized. Use the rubber ball gun to stun any problem makers. Of course, we don't want any of this to transpire," Gunner finished.

The four—Gunner, Dale, Carrie, and Pierre — walked out of the tree line in which they had been hiding, carrying the deer. They were calling it a deer, anyway.

The hominid group consisted of twelve adults and three smaller individuals, "children," who looked to be six or eight years old. Two of the males, the largest of the group, well over 1.9 meters in height, stood up and moved away from the group that had been sitting, working on various tasks. Two other males also stood up and put themselves between the first two males and the remainder of the group. All the males held spears in their hands, which they raised, ready to be thrown.

Dale and Pierre set the deer down on the ground. Each of the four members of the landing party pointed at the deer and then pointed at the natives, seeming to indicate that the animal was a gift. They took several steps backward, all the while facing the natives. They held out their arms to their side with their palms opened, as they bowed their heads slightly. What happened next stunned everyone.

The two closest males dropped their spears and knelt on one knee with their heads bowed. The other members of the group did the same. Dale and Pierre stepped forward to the deer carcass; they picked it up and carried it over to the larger group that was still kneeling. Dale and Pierre set it down.

"Greetings, I am Dale," he said as he pointed to himself.

Pierre also said his name while he pointed to himself.

Dale pointed to the largest male and asked, "Name?"

The tall male laughed and then pointed to a redheaded female and said, "Akeye."

Dale looked at the female and said, "Greetings, Akeye, we are from Earth." He gestured to his group and then pointed at the sky. Akeye walked over to Dale, held out her arms, and then embraced him then Pierre. She looked over to the other two, Gunner and Carrie, smiled, and waved at them.

"Maa, nung lowng," Akeye said, indicating that they should come and sit down.

Gunner looked at each of his group. "Let us sit down and join them." Next, he spoke into his arm comm device. "Penny, hold at your position until instructed otherwise."

As they sat down, the natives built a fire and two of the men started cutting pieces of meat from the already dressed deer. Akeye went over to the pile, took a piece of skinned meat, and bit into it, ripping a big bite off with surprisingly strong and sharp teeth that became visible for a moment. She smiled wide, revealing the blood from the raw meat and handed the piece to Dale who turned green. Dale managed a weak smile as he held the bloody piece in one hand and timidly bit off a small bite. After what seemed like an eternity of chewing, Dale tried to execute a swallow but was entirely unsuccessful. The meat started to go down the hatch, but the hatch had other ideas as Dale heaved and projectile vomited the much-chewed piece of alien animal flesh into the fire.

Everyone seemed to pause for a moment; then the natives and the *Calypso* crew all burst out laughing at Dale, who sat still for a moment and then proclaimed, "I feel better now."

He reached for his pocketknife and fished out the smoldering piece of animal flesh and bit into it again with much better results.

"Well, Dale, looks like our lesson on cooking meat and other food got an early start," Gunner said as the *Calypso* crew laughed again.

Dale invited the Akeye clan to follow them back to the shuttle after they finished "lunch." They were only about five kilometers away from the shuttle, so they reached it in an hour. The clan, which they were now referring to as the "Akeye clan," was very organized in their order of march. Within five minutes, they had gathered up everything they had with them. Each person carried something, even the children.

As they drew near the shuttle, Gunner whispered to Dale, "Are they going to freak out when they see the shuttle?"

“I expect they will be very curious and not understand what it is. Did you get in touch with Hatcher?” Dale asked.

“Yes. He is getting the portable habs out. We’ll see if they want to use them. They are easy to set up, take down, and carry. Hopefully, they will see that as something that can make their lives easier.”

“The one thing we do not want to do is to have them see our weapons. We don’t want to upset the balance of power that may exist between clans. Peaceful cooperation between groups is our first objective,” Dale said.

When they came out of the tree line, the shuttle became visible to the group. The lead male, who was walking with Dale and Gunner, motioned for all of them to stop as he went back to Akeye. They spoke in their tongue for a moment and then finished. The lead male, who had told Gunner that his name was Keye, came back to Gunner and Dale and waved his hand from side to side while shaking his head. Dale took this to mean that they did not want to approach the shuttle.

Gunner got Hatcher on his comm device and said, “Hatcher, I need you to come out of the shuttle, waving and smiling at the clan members. We need them to know we aren’t frightened of the shuttle.”

Hatcher descended to the lower level of the shuttle and walked out of it while waving and smiling. Gunner and the other four smiled and waved back. Dale turned to Keye and smiled, gesturing to him and the other clan members to follow them up the hill.

When they arrived at the shuttle, Hatcher and those who remained behind were there to greet them.

"I want to show the Terrans these hab units. I've got one still in its pack," Hatcher said as he set the pack on the ground and pushed the AUTO button.

The contents of the pack unfolded on their own and began to "assemble" themselves. First, a floor was formed, then walls, a door, and finally a roof. The overall dimensions were three by four by two and one-half meters: sufficient room for a family of four, but certainly not a luxurious five-star accommodation.

It was at this point that it began to sprinkle lightly. A moment later, the sprinkling turned into a downpour. Gunner motioned for the clan to seek shelter in the habs, which they gladly did as the rain and wind were quite strong.

As the shuttle crew ran to shelter inside the shuttle, Gunner asked Hatcher, "What did you call these people?"

"Terrans. This is Terra Nova, so they are Terrans, right?" Hatcher asked somewhat flippantly.

"So they shall be called," Gunner answered back.

CHAPTER 82

Settlement Plan

THE LANDING PARTIES returned to the *Calypso* at the end of their weeklong stay on the surface. A meeting of the settlement committee was held the day after they returned to the ship, chaired by Dr. Hans Larsen.

"We've been going over the data in real time. As a result, the staff on *Calypso* are well aware of the findings of each landing team. We mourn the loss of Dr. Ben Shaparo, who was trampled to death by wild elephant-like animals at landing zone number three. This is a sober reminder to all of us that the job in front of us is inherently dangerous. Ben will be remembered as a gentleman and a first-rate scientist. He is the first human to die on Terra Nova and will not be the last, unfortunately," Hans said.

"Well, I guess that's one way to shut Ben up," Gunner whispered to Simi who shot a disapproving look back at him.

"After reviewing the data, the settlement committee has decided to make our initial settlement at landing zone number one. Gunner and his team were able to easily locate Terran hominids and contacted them with good results. The location is approximately forty degrees north latitude on the southeast coast of the largest northern hemisphere continent. We believe the winters are moderated by the proximity of the ocean, which also makes for moderate summers. Most Earth fruits and vegetables will grow in this location, minus tropical ones. We will cultivate Earth crops immediately to help ease us into the introduction of native Terran crops into our diets.

"We have prepared a draft settlement plan for this location. It is available for all of you to read on the ship's common drive. The settlement construction and security force are in the first group of seventy-five colonists. They will be putting in shelters, power, water, and sewage systems for the colony. They will need six weeks to get enough of that infrastructure in so that colonists can be brought down. I know everyone wants to go down right away, but we will have to keep building shelters and continue expanding the power, water, and sewage systems.

"Once the initial phase of infrastructure is in place, approximately one hundred colonists per week will be going down. This will take more than six months Earth time. We will be developing a departure list that will be posted before any colonists are to go down. We've decided to name our first settlement Benston, in memory of Dr. Ben Shaparo, who will be buried at that location when time permits. Elder Spencer Oaks, our nondenominational Unitarian Universalist minister, will be making those arrangements. Are there any questions?" Hans asked the assembled committee. At the end of a three-hour-long meeting, no one had any questions.

"Are you going to come down right away?" Hatcher asked.

"Dr. Larsen will determine the order of when everyone goes down," Terra answered.

"I just thought you might have knowledge of what Dr. Larsen had developed," Hatcher said sheepishly.

"No, I'm not putting myself ahead of others," Terra said.

"Speaking in those terms, my dear, how will we be governed on Terra Nova? I know that everyone here looks at you as Eve's representative. Will you be a 'queen'?" Hatcher smiled as he asked.

"No!" Terra responded.

"You are already my queen," Hatcher said as he leaned over to kiss Terra deeply.

Terra stared at the wall as she hugged Hatcher, thinking about what he had asked. *How will we be governed here?*

She knew the *Calypso* Charter made clear the senior captain was the mission leader. *But how should things run now that they were here?*

CHAPTER 83

The Terran Compact

TERRA ENDED UP coming down with her parents and twin sisters almost five months later. This was for the best as there were many hazards in the new world, and toddlers had a knack for finding hazards before anyone else did.

The shelters for the colonists were durable, albeit spartan domiciles. The shelter designers had tried to mitigate this with assorted color patterns on the exterior and interior. The structures themselves were made from a material that wouldn't burn, was well insulating, and was guaranteed for three centuries.

Sam was skeptical about the last claim. *These things weren't even around more than twenty-five years ago. Maybe some people believed in the use of accelerated aging tests, but I don't. I've seen too much highly rated comm equipment fail in half the guaranteed time. But it would probably last for a long time*, he thought, as they brought the last of their belongings in.

The structure had three bedrooms, a kitchen, and a universal room that could be used for many purposes. Showers and toilets were shared among two dwellings in what was called "the bathhouse." The house was basic, but the rent was more than acceptable. Each colonist was assigned work by the worker's council based on their age, education, and background. The supply council ensured that each person had clothes, food, and medical care as needed. The food stores were sufficient to keep the entire colony alive for three years, but everyone was more than anxious to grow fresh vegetables and fruit after only a few days on the store's rations. The palatability of the food stores was the subject of jokes for generations afterward, but no one starved, not during those early years anyway.

Captain Simmons, who was still acting as the leader of the colonists, had asked Terra to form a committee to draw up a charter on the new government that would be used to govern the affairs of the colonists in the new colony. Terra had five academics that she tapped to help draw up the charter and a bill of rights for the colonists. They met twice weekly to go through the progress of the draft charter.

"Dr. Sue Stein has the introductory section, which includes the rights of citizens. She is going to take the balance of our time tonight to present it," Terra said after the meeting was called to order.

"I have put the draft on the protected colony drive. I hope you all had a chance to review it on your colony comm device. I've listed out the rights of each citizen in bullets in the second paragraph. Most of you will recognize that these rights are the same ones listed in the FBS Charter of 2129. Those, of course, were lifted from the new US Constitution of 2047. They should look familiar to one and all. There's no need to reinvent the wheel when the wheels we had worked so well for centuries, right?" Sue asked rhetorically.

"I think we can probably take most of our constitution from the FBS Charter," Manuel Esquivel, a human rights lawyer, noted. "I mean, what didn't we like from the charter that we all lived under? It was truly a representative democracy blueprint. It didn't have structural flaws like the US Electoral College or their old Senate. It protected against legalized bribery, also a feature, not a bug, of the old US Constitution."

"But what it didn't do is check sociopathic ambition. We still had people whose sole aim in life was the accumulation of power and wealth," Terra interjected. They all turned to her at this point, surprised that she had said anything, which she had not done in all the previous meetings.

She continued. "Furthermore, we still had unqualified, and sometimes deeply flawed, individuals who ascended to high office. It almost seems like to me, anyone who vigorously aspires to high office is, perhaps, deeply flawed. Most people want to live their lives in peace. They don't desire power, position, or money. Yet they live at the whim and will of defective souls who desire to control them and hoard power and wealth. Perhaps there is a better system, at least for how people are chosen for positions of power and public trust?"

For a moment, they all remained silent, thinking about what this young woman had told them, which to all of them seemed to cut right to the heart of the matter.

Carla Chan—Peter's wife, a statistician, and political scientist—spoke. "We've always assumed that elections were the one and only way for leaders to be picked, but as Terra noted, this often had disastrous results. Think of the early twenty-first-century president, popularly referred to as the Traitor. Consider also the dangerous political party that lined up behind him and continued after his death in prison. This group of people and the leaders they chose unleashed xenophobia, racism, factual denialism, and political violence that caused the second US civil war. On this planet, two intelligent species are going to live together for a long time. This alone is a recipe for the revival of the old racist tendencies that some believe are inherent in our genome.

"What if we prescreened or prequalified a large group of our citizens with the intent to develop a pool of people from which leaders could be randomly selected? Because they would be prescreened for desirable attributes, they could govern much more effectively, precisely because they did not have the racist, sociopathic, and whatever other tendencies that we know leaders should not have. We could ensure that this pool of people had the intellectual capacity to handle the myriad complex issues that seem to come up in the course of governance. We could then devise a random selection mechanism that would pick from this very large pool for the positions of trust and leadership. The makeup of our leaders could also demographically represent the public whom they serve. We wouldn't have just wealthy individuals or lawyers." There was nervous laughter from the two lawyers present. "Instead, they would be engineers, scientists, domestic specialists, academics, and even blue-collar workers who, of course, possessed the native intellectual skills to govern. We would limit the amount of time in office in a way that balances the need for competency in office while protecting against the professional office holder."

Everyone was quiet for several moments.

Finally, Terra broke the silence. "Thank you, Carla. Well said, insightful. I was looking and hoping for this kind of out-of-the-box thinking. We have an important job to perform. I think we should spend the next few days before our next meeting thinking about what Carla suggested."

In the successive six meetings, the group hammered out what became known as the Terran Compact. Acceptance of the document was subject to a vote of all the colonists above the age of sixteen. On the appointed day, March 31, 2353, the Compact was accepted by an astonishing 94 percent of all eligible voters present.

For the next six months, the committee on public service defined the parameters that constituted a colonist who was considered "QUALIFIED FOR PUBLIC SERVICE." The first random selection of a Terran president and congress (only ten members given the small size of the colony) was made on November 1, 2353, with a tenure of service that was limited to eight years for the congress and ten years for the president. An engineer named Fisher Davis was selected as the first president and was commissioned for duty by Terra Nova Winstead, representative of The Mothers and Mother of the Terran Peoples. A day of celebration was enacted every November 1 thereafter to memorialize the events of this first selection day.

CHAPTER 84

The Life of The Mother

"OUCH!" TERRA SHOUTED softly as her hammer gun pushed down on her thumb instead of the nail that it was supposed to push into the lumber. The colony had started building with native Terran materials as the prefabricated housing units had run out and it became clear that more housing was needed, given the cramped conditions in so many housing units.

"Are you OK, Terra?" Hatcher asked.

"It's just my thumb," Terra answered with a smile. "I've still got one good one. I'm fine."

"You look really tired. I know you didn't eat much this morning. Are you really feeling OK?" Hatcher asked with genuine concern bleeding through his words.

Hatcher didn't know it, but Terra had taken a pregnancy test that morning. She wanted to work that day as the need for additional housing was so great that she felt like it required the work of everyone to alleviate it. If Hatcher knew she was pregnant, he would have made a big fuss about her taking it easy, which she didn't want to do.

Hatcher walked over to her and reached from behind her to hug her. At that very instant, Terra could no longer suppress the need to empty her already empty stomach even more, and she vomited onto the roof on which they were working. He could see she was not steady on her feet. Hatcher grabbed her to steady her. She sat down, suddenly quite dizzy.

"Terra, is there anything you need to tell me?" he asked, smiling widely.

"Well, I was going to wait until this evening, but I don't think I can hide this much longer. Hatcher, I'm a little bit pregnant," she said weakly.

"Wait, what? Really? Oh, Terra, this is fantastic! You should get off this roof right now and go home and lie down, get some rest. I can finish this," he said with what came across as a clearly patriarchal tone.

"Maybe so. I'm going to need help getting down off this roof though. I'm dizzy," she said as she slumped over and fainted on the roof in the warm spring sunshine.

Hatcher stood and shouted, "I need help, Terra has fainted! Can I get some help to get her off this roof?" Four men working on a housing unit next to the one Terra and Hatcher were on ran over to the ladder leading up to the roof.

"Be careful," Hatcher said as a giant native Terran picked her up, put Terra over his shoulder, and climbed down the ladder.

"Is Terra OK? I'll go get Dr. Harrison!" one of the other men said.

"She's OK. She's just, well, she's pregnant," Hatcher said, somewhat embarrassed.

"What was that? Terra's pregnant?" asked Gunner, who was now among an even larger group of colonists who were streaming over and surrounding Terra.

The ensuing babble of voices became even louder as the word spread that Terra was pregnant. For his part, Hatcher tried to play down the excitement as he knew his wife did not like to be the center of attention. Simi, Gunner's wife, came over to Terra and laid the palm of her hand on Terra's abdomen, saying, "Be calm, be still, . . . peace." Terra opened her eyes and stared at the assembled crowd.

"Hatcher, what's going on?" Terra asked, mortified.

"I needed some help to get you down off that roof. Anyway, I don't think you were ever going to keep your pregnancy a secret. They all know now," Hatcher said as he turned abashedly toward the crowd of people who spontaneously started clapping, smiling, and congratulating the young couple.

A short while later, they sat with Dr. Harrison as he looked over e-scans. "Looks like twins, a boy and a girl. You're only five weeks along. Since you are dehydrated and dizzy, I'm going to give you some IV fluids," Doc Harrison said in his slow southern drawl. "I want you to stay in or around this house until you start eating and drinking more without nausea. This ole country doctor has delivered more than a few babies, and I know what I'm doing. No ifs, ands, or buts—you take it easy!" Doc Harrison closed his medical bag and excused himself.

Hatcher, Terra, and her parents sat quiet and motionless as Terra's toddler twin sisters cooed on her parents' laps.

"Well, don't you even worry about the housework. I got it, Terra. Congratulations, honey, you'll be such a good little mother!" Riley said as they all broke the silence with smiles and laughter.

PART THREE

Lost in the Past and in Space-Time

CHAPTER 85

Farewell *Calypso*

SINCE ITS ARRIVAL at Terra Nova, the *Calypso* had slowly been emptied of its huge cargo load. The four shuttles were in a constant mode of leaving, landing, unloading, and returning for more loading and starting the process over again. Technicians from Preemus ended up traveling to Terra Nova to install the additional hibernation pods for the Mars colonists and resupplying failed equipment on the *Calypso*. They also relieved the colonists on Terra Nova of the need to replenish the *Calypso*'s food stores, as well as seeds and plants for the *Calypso*'s internal fresh food production.

Five weeks after Terra's happy news broke wide open in the colony, the *Calypso* was ready for departure. The ship would leave a shuttle behind on Terra Nova so that the colonists could conduct long-range reconnaissance of the landmasses to assess hominid evolution in other far-flung locations. Communication with the Mars colony was still intact via EPC channels, and it was turning out to be a dicey situation. Mars was hanging on by making do with many equipment items whose service life had long expired. It turned out that those who designed the life-support systems had overdesigned the equipment, which was very fortunate.

The return crew was manned by volunteers who wanted to return and help in the rescue of the Mars colonists. Captain Simmons and her husband, Timeye, were going on the return trip because of their experience gained en route to Terra Nova. They knew how to use The Mother's map system for the universe, which allowed for much shorter travel times using wrinkles in space-time. Gunner, Simi, and their son, Gunner Jr., changed their minds and opted to stay on Terra Nova. Peter Chan, his wife Carla, and their two children also opted to stay. Stuart and his brother Carl were returning with Captain Simmons. They were both bachelors and had no children. Both were interested in mapping out star systems that they would see on the return route and so their choice was an easy one.

Terra's friend Cassia also wanted to return and help with the rescue. The returning crew would not see their crewmates again for almost eighteen years. Given that human life expectancy was now just over one hundred forty years, this did not mean that they would never see their associates again. They would be older, but not geriatric. It also helped that relativistic effects would shave almost a year off the amount of time for the returning crew.

President Fisher Davis, his wife Rhonda, Terra, Hatcher, Gunner, Sam and Riley, and a handful of family members of the departing crew traveled up to the *Calypso*, still in orbit, to say goodbye to the crew.

"I'll try not to get emotional about this," Captain Simmons said, wiping a tear from her eye. "But it is so hard to leave you all here. We have spent years together, and for a time that will come to an end. But when we return, we will bring so many new friends with us. I want to emphasize many because it will be over twenty thousand. I think you better start now on getting the homecoming dinner ready," she said as everyone laughed with her.

"Sandra," Terra spoke, shedding a tear. "I literally grew up with you on this ship. You were like another mom to me. Later, you were my commanding officer. I think I liked it better when you were my other mom! I am going to miss you. I will miss all of you. You are all embarking on another long mission, an especially important one, to rescue our fellow Earthlings on Mars and some on the Earth as well. May The Mother bless your voyage and those on Mars who await your return and rescue. Farewell, *Calypso*, our mighty ship. Bring our friends safely home again." Everyone assembled in the shuttle hangar bay clapped.

Twelve hours later, the *Calypso* broke orbit and was on its way to Mars.

CHAPTER 86

FBS *Calypso*, Outside of the Mu Arae Heliosphere

SIX MONTHS INTO the return voyage, Mu Arae was fading quickly, though it was still the brightest star in the heavens by far. Stephanie was preparing for the first wrinkle transit with Dieter's help. The first transit would be followed almost immediately by a second, and she wanted to have everything planned for both before the time came.

"How is it coming with checking out the software link between The Mother's map and our guidance and control computer?" Stephanie asked Dieter.

"It's finished," Dieter answered. "I've run some trials, but it seems to be working just like it was for us on the outbound trip."

"Well, I just don't want to end up in the M83 galaxy by mistake."

"I thought that was one of your favorite twenty-first-century music groups?" Dieter asked with a chuckle.

"Yeah, I love the electronic-style sound for which they were famous. I liked Outro the most. It stirred my soul!"

"I can't wait to get some good Mars salsa," Dieter said.

"Well, it's going to be a while before you can taste some."

He sighed and said, "I know. One of my favorite twentieth-century songs by a totally hot babe goes:

'Anticipation, anticipation, /

Is makin' me late, /

Is keepin' me waiting.'

I guess I can't do anything but 'wait' and 'anticipate.' How about we go back to my quarters and rock out to late twentieth and twenty-first century tunes after Anna gets here, which is, wow, only ten minutes from now! Time flies when you're having fun." Dieter smiled at Stephanie.

Stephanie smiled at him and said, "OK, babe, but let's make sure we stick to music. You wore me out last night!"

Anna Stetson showed up, early as always, which was a feature of her personality, not a bug, Dieter noted silently.

"Hey, Steph, how is it going on the planning for the first jump?" Anna, always bubbly, asked with her trademark ear-to-ear grin.

"All done. The file is open, please check it for me," Stephanie said as she winked at Dieter.

"Okey dokey," Anna answered, smiling again as Dieter and Stephanie left the room.

CHAPTER 87

Interstellar Space

THE *CALYPSO* HAD reached 4 percent of the speed of light, and it looked like they would reach 5 percent before the deceleration phase, which was now scheduled to begin in another six years. The additional efficiencies that The Mother's engineers were able to add to the electromagnets in the ion accelerator had worked even better than anyone expected. Of course, this meant that the flight plan had to be revised three times now. The extra velocity meant they reached the wrinkles sooner than expected, and their destination, which like everything else in space was a moving target, had not moved as far as expected. This was all good news, and it had been shared with Terra Nova and the Mars colony.

It couldn't have been better news for the colonists on Mars. They were doing OK, but things were wearing out and breaking down. Additionally, they had received some communication from Earth, which meant there were still survivors there that would need to be picked up.

The Mother's people had added hibernation capacity beyond the original plans in case more survivors turned up on Earth. Captain Simmons was not sure what to do if there were more survivors than hibernation tubes. At the moment, she assumed she might have to leave people behind, but she did not want to leave them helpless. She had ordered that food production be increased and freeze-dried. If she couldn't take them, she wanted to leave them with food and other supplies.

Captain Simmons was three months along. This was her first child, and the nausea that the pregnancy caused made her wonder why she thought this was a good idea. As she flushed the toilet and watched her breakfast disappear down the drain, she thought about why all the many remarkable medical breakthroughs of the last several centuries had not resolved the issue of morning sickness.

Probably because men viewed it as a minor inconvenience. And now that gestation could take place in birthing tanks, no one considered it a vital concern, she thought.

The *Calypso*'s birthing tanks had been transferred to the surface of Terra Nova, though the new trend was to experience 'natural gestation', as had been done for thousands of years.

Pity they could not have left a few birthing tubes on board, she thought.

* * *

Timeye was studying the original plans for the ion drive electromagnet accelerator assembly. In doing so, he started to see the origin of the improvements The Mother's engineers had made to the ion drive. Using the enormous output of the propulsion nuclear reactor and adding an additional acceleration array greatly increased the thrust of the ion propulsion system.

The additional array, or at least the first array the electrons were exposed to, accelerated them at higher velocities than the original one-stage accelerator. This meant that by the time they were shot at xenon atoms, they were traveling faster and colliding with more xenon atoms than the original design. Now, rebounding collisions started playing a bigger role in creating more and more positively charged xenon ions. More mass meant more thrust, faster acceleration, and a greater velocity at the time they were ejected into the nozzle assemblies. Nothing really revolutionary, but not something Earth designers saw fit to risk in the original design. Bottom line, Timeye concluded, they were going to get back to the Mars colony quicker than planned, which meant that they would get "home" to Terra Nova sooner.

Timeye wanted their daughter to experience her teenage years with other teens on Terra Nova. Even a man who prided himself on logic and rational thinking knew that emotional maturity and well-being were enhanced by experience with other children. He sometimes wondered if it was the best idea when he and Sandra had volunteered to go back for the rescue of the Martian colonists. In the end, they decided that their knowledge of the route and the possible dangers would improve the probability of mission success. It was their duty, and Caldonians were all about their duty.

When Timeye got back to their quarters, he found his pregnant wife sucking on a popsicle in the bathtub.

"Are you OK?" he asked.

"My stomach doesn't seem to want to keep down anything except this," she said, holding up a popsicle stick and exposing bright green lips from the coloring in the popsicle.

"Your lips . . . what happened?" he asked with obvious concern.

"I guess you've never had a popsicle, Timeye?" she answered with a smile and a hearty laugh. "You should try one. They're really pretty good. I like the green ones. I've had three of them!"

Timeye smiled and nodded, understanding now the cause of the green lips on his wife.

"Kiss me!" she said.

"I don't know if I should kiss my commanding officer while on duty," Timeye answered with a smile and desire in his eyes.

"I won't tell anyone," Sandra said as she rose from the bubble bath, exposing her growing breasts to him.

This was just too much to resist, even for a man of duty and science. Timeye pulled his clothes off and stood in the large tub in front of her. He kissed her gently then with more fervor and wrapped his arms around her at the waist. He pulled her closer, and she moaned softly as he did.

"Your duty can wait for a few minutes, sweetie," Sandra said as they stepped out of the tub and headed for their bedroom, dripping all the way.

CHAPTER 88

The Oppositeverse

THE ALARM STARTED sounding at 4:32 a.m. Sandra pulled her uniform on and started running down the hallway to the guidance and control section with Timeye right behind her. When they got there, Anna Stetson and Stephanie were hunched over the screens on the panels in front of them.

"What's going on?" Sandra asked.

"We were transiting the sixth wrinkle when everything started making no sense!" Anna answered.

"Please, Ms. Stetson, try to be more precise. What do you mean?" the captain asked.

"Well, we seem to have entered a region of space where the rules operate opposite of what they should," Stephanie said. "For instance, that strange star directly in front of us is pushing us away."

The captain thought for a moment and said, "Yes, I see. It should be pulling us toward it, but it isn't."

"Yeah, well, it is pushing us away, but that is not the only thing. All of our electrons in the propulsion system are now positrons, positively charged, and that means we're not generating thrust. No electrons, no collisions producing xenon positively charge xenon ions. In fact, our electromagnet assembly shut itself down because of the buildup of positive charge in the gun assembly," Anna said, having completely lost her cheerful disposition and smile.

"We are slowing down from the forward push against us by that star and by the loss of our propulsion thrust. We will begin moving in the opposite direction by noon today unless we do something or figure something out," Stephanie added, exasperated.

As the captain pondered what she had been told, Timeye turned to Stuart, who had arrived unnoticed in time to hear the discussion.

"Stuart, what is your take on this situation?" Timeye asked.

Stuart didn't answer Timeye immediately; instead, he asked Anna and Stephanie, "Did this begin to happen *after* we pushed through the wrinkle?"

"Yes," they both answered together.

"Anything else anomalous?" he asked.

"Well, we seemed to accelerate to 50 percent light speed as we slipped from one side of the wrinkle to the other. But that was clearly some instrumentation error," Stephanie said.

"I mean, we would have all been flattened like a bug on a windshield had we actually accelerated that much," Anna said, now with a hint of a smile on her face as she laughed at her own analogy.

"Yes, but we all did experience something, didn't we?" Stuart said, with an ashen look on his face.

Captain Simmons, Timeye, Stephanie, and Anna looked at Stuart with a puzzled look.

"Stuart, what's going on with your hair?" Anna asked.

"Yes, my hair seems to have regained its foothold on my head, not only that, but I don't have much gray hair left either," Stuart answered.

Everyone stared at Stuart's head, which had always been shaved bald to cover up the fact that he had already lost most of it years earlier.

"Stuart, what does this mean? Electrons becoming positively charged, a star with a repulsive force of, what, anti-gravity? You getting hair back on your head. It's like we are in the mirror opposite universe to ours, with the rules of physics reversed, including entropy . . . You know, all of you look younger!" the captain said with a gasp.

In fact, Captain Simmons was correct. She looked at Timeye and noticed he did not have gray hair anymore.

"I don't mind getting younger, but at this rate, we have maybe just hours until we are all children again," Stuart said soberly.

"I'm just guessing here, but I believe we should let that star push us backward back across the wrinkle. Maybe that will put us back into our universe?" Timeye proposed.

"OK, but we will need to steer ourselves as we head backward. Our course had us on a curved trajectory coming out of the wrinkle. But wait, if we want to turn left, do I need to pretend I need to turn right? I mean, what things, exactly, are reversed?" Stephanie asked, concerned about how to steer the ship.

"I believe that will not be necessary. *If* we are in a universe that is the mirror opposite of ours, the fundamental charges, fields, forces, and entropy will be operating with opposite signs. What direction we go is not a fundamental quantity. I think we should assume we steer back along the arc that we traversed since leaving the wrinkle," Timeye said.

"Have you noticed that there are almost no stars in the star field? I see a lot of gas, nebula gas, but very few stars. It's almost like I am looking at a very young universe. If this universe is a mirror of ours, if we stayed around for a couple of billion years, we might get to see the Big Bang in reverse," Stuart said, chuckling, having regained his composure.

"Personally," Stuart said, "I'm hoping my youthful looks return and leave me at a younger age."

"OK, folks, let's work the problem. Stephanie, I want you, Anna, Stuart, and Timeye to work through our backward course. I've got to tell the crew something. I don't know what other oddities we may discover in the next few hours. Will my . . . wait! Will my baby grow in reverse? Oh God!" the captain exclaimed. Timeye came over to the captain and hugged her, holding her tightly.

"It will be OK. When we cross back over, perhaps the baby's growth will reverse. Maybe you'll just be pregnant for a longer time?" Timeye said hopefully.

CHAPTER 89

Now Where and When Are We?

BY 10:00 A.M., they had the ship turned around and had reversed the polarity of the electromagnet assembly and restarted it, using positively charged electrons or positrons to bombard the clouds of xenon atoms to produce ions with a net charge of opposing sign. The strange electrons worked as well as negatively charged ones, especially now that the electromagnets would move them in the manner they expected. The ship stopped decelerating and began accelerating toward the wrinkle.

Stuart determined that the wrinkle they traversed was not like the one that The Mother's map indicated. Instead, it was a much, much bigger wrinkle, and that had accelerated the ship to 50 percent light speed. Reversing course seemed to be doing the same thing: the ship was now accelerating toward the same speed, but the odd thing was no one seemed to notice a crushing acceleration.

At a point where Stuart predicted they would have moved back to their starting position before they entered the wrinkle, Stuart surveyed the star field. There were now many stars visible, but he did not recognize any of them. He had no idea where they were now. Locating known stars from their brightness would work, but it would take time.

Stuart said, "Captain, this will take some time to figure out where we are, and we must do that before we can map a new course back to our solar system. In the meantime, we are accelerating at something like 50 percent of the speed of light, so we are going somewhere very fast. If we can aim ourselves in the right direction, we could make very good time."

As he finished, Stephanie and Anna came in, excited, with seemingly good news.

"We've found three known pulsars. The guidance computer used these three known stars to figure out where Earth is and where we are. It seems that we are not very far from Earth, but now we are one hundred eighty degrees opposite of our sun," Anna said. "It's like we reentered our space-time at a different location. We overshot our destination, but at least we don't appear to be very far from it, not as far as we were before we entered the wrinkle."

"OK, but do we have a new flight plan and a deceleration plan? We've got a lot of velocity that we need to mitigate. Otherwise, we'll just wave at the Mars colony as we rocket by them," the captain said.

"We're working through that. We have a couple of initial options, but I wanted the computer to work through all the options so we can choose the best one. We'll have it figured out in a couple of hours," Stephanie answered.

A couple of hours later that evening, Stephanie and Anna brought Stuart and the captain new course and braking options. The new course would have them at the Mars colony by December of that year, seven months hence.

"We can be in orbit at the Mars colony in seven months, four years sooner than we expected originally. Perhaps between now and then we can study what happened to us and utilize the same wrinkle to return quicker or not. I mean, we don't really know what happened to us. Right now, I'd be reticent about hoping to end up in a similarly favorable situation," Stephanie said thoughtfully.

"Agreed. Stuart, can you and your team keep working to figure out what happened and if we can exploit this phenomenon?" the captain asked.

"Yes, we will work on it. I really don't understand what happened, and I don't like that. I want to know what happened."

"Anything else? It's been an eventful day," the captain asked.

"Captain, there is one more thing," Timeye said. "Since we are traveling at 50 percent the speed of light, not 5 percent, there is going to be a significant difference between our clocks and the clocks on Terra Nova and Mars. We will experience significantly less time than either location. Furthermore, we will be traveling at a much higher velocity for the remainder of the voyage. We will experience even more slippage for the next seven months. I noticed we missed our daily comm cycle with Terra Nova and Mars. When we reestablish comms, we should ask them what their date and time is. When we missed our call to them, why did they not call us? Does anyone have an answer for that?" Timeye asked.

Everyone looked at one another with blank stares. An answer was not forthcoming.

CHAPTER 90

No One Picks Up

EARLY THE NEXT morning, Captain Simmons came to ComCent to be present for the morning call in to Terra Nova and Mars. Normally, this involved a short message on their position and ship status along with any specific questions that those on the receiving end might provide help on. An acknowledgment of receipt of the message would follow, along with a status on the colony or other relevant information. Answers to questions often required time to get, and they would be sent in the next comm cycle or later. Today's message was sent with an explanation of the events of yesterday and with questions asking about the date and time at each location. This normally would be followed by an immediate response acknowledging receipt of the message. As the minutes passed with no message received, everyone got restless.

"Why would they both be taking this long to acknowledge the message?" Stephanie asked the group gathered around the EPC console.

"Don't they normally respond quickly?" Stuart asked Sal Espinoza, the on-duty comm engineer.

"Yes, usually the acknowledgment is sent within a couple of minutes."

"Has there ever been a failure to acknowledge a message?" Captain Simmons asked.

"There was a day, about a year ago, where we had a coolant system problem, and the entangled particles started to lose their supercooled environmental conditions. That can happen, and it can cause a loss of particle stability. We weren't able to acknowledge their message back to them for several hours. When we finally acknowledged their message, it was clear that they were rattled by our inability to answer immediately. I guess it's our turn to be rattled," Sal said with a faint smile.

Stuart was silent for a moment then spoke. "Perhaps there is a very obvious reason why they are not answering."

"Do share, Stuart," Stephanie said.

"They aren't *there yet*," he answered.

"What do you mean, Stu?" Captain Simmons asked, clearly puzzled by his response.

"There was a theory, put forth in the late twenty-second century by a mathematician named Roberto Brando. This theory postulated how velocities greater than the speed of light could be achieved *and* what the consequences of traveling at those speeds would be."

"But we can't travel faster than the speed of light," Sal said, a bit unsure of himself.

"Well, that is not exactly and strictly so. The wrinkle that we traversed then reversed back out of was much, much steeper than The Mother's map indicated. I've been doing some calculations on what our *actual velocity* was when we traversed between the sides of the wrinkle. It was much greater than we thought. Much greater than the speed of light. In space-time, nothing can go faster than light, but that is not the case if you are *outside* of space-time. When we previously transited wrinkles, the wrinkles were relatively small. We enter the side of each wrinkle normal (perpendicular) to the surface of the local space-time and punch right through, minimizing our distance through no-man's-land," he explained.

"What is 'no-man's-land'?" Stephanie asked.

Stuart paused then answered, "It is no longer in our space-time. Where it is, what the 'rules' there are, and what limitations there are on velocity, I am not sure. When we exited the wrinkle, I reviewed the captured star field image. The North Star is not where it is supposed to be. It is known that over great lengths of time, that star's position is not fixed. It has appeared to be fixed over time, but it was not always a constant indicator of direction as we know it today. Thousands of years ago, other stars were used to provide indicators of direction for ancient people. Ancient civilizations around the Mediterranean Sea built structures that had alignments to where known stars were at the time they were built. This was our first clue that the heavens, in fact, do move over long periods of time, thousands, tens of thousands of years. The current position of the North Star is where it would have been about twelve thousand three hundred years ago. We have traveled back in time. Mars had no colony twelve thousand or so years ago. There were no Earth people on Terra Nova twelve thousand three hundred years ago. They didn't answer us because they aren't even alive yet."

"Stuart, you definitely got our attention, but I must ask, can we ever get back to our time?" Captain Simmons asked.

Stuart did not answer immediately. He paused, looked at the floor, then answered in a measured manner, "The fact that we went back in time, which no one ever believed was possible, leaves me hopeful that we can reverse the situation though at this time, I don't understand the principles well enough to know what to do. Perhaps in time, I can figure it out. It would help if we could get assistance from The Mother. She might know how to make this work."

"Was she, Eve, alive twelve thousand years ago?" Captain Simmons asked.

"When we visited her, I recall her story in some detail. She indicated that their trek back to East Africa came as a result of the Younger Dryas Cooling Event. That was eleven or twelve thousand years ago, as I recall. So I think she would be alive if I'm recalling things correctly," Stuart answered.

"If we return to Earth, the Earth of twelve thousand three hundred years ago, maybe we can find her and she could help us?" Stephanie asked out loud.

"Perhaps, but would we know where to even look for her?" Captain Simmons asked.

"The Great Cracked Valley, we might be able to find that. I guess I assumed that was the Great Rift Valley in the old African country of Kenya. We might actually be able to find her," Stuart answered.

"So let's set a course for Earth, and go talk to the younger Eve of Eden, our 'Mother,'" Captain Simmons announced.

CHAPTER 91

We Were Expecting You

CAPTAIN SIMMONS'S PREGNANCY did continue, although her term ended up being eleven months. Just before they established orbit around the ancient Earth, she gave birth to a son, which Timeye and her named Erik. The child had bright-red hair, similar in color to Timeye's hair when he was young. Captain Simmons turned command of the *Calypso* over to Timeye while she convalesced with her newborn son.

The crew was able to slow the Calypso down by making two passes through the Jovian atmosphere which resulted in the ship being able to successfully slow down and reach the Earth. As they approached the Earth, Timeye and Stuart organized a landing party comprised of Stuart, Dale LaPlant, Edwin Godwin, an archaeologist, Stephanie, Dieter, and Camilla Prescot, a ship security officer. Stuart was in charge of the landing party, and Stephanie piloted the shuttle.

"The Mother recounts that they descended to a large lake in the Great Cracked Valley," Edwin noted and continued. "Lake Victoria, the largest of the African Great Lakes, forms a part of the headwaters of the Nile River system. The Victoria Nile drains from the northern end of the lake and makes its way to the Nile River proper. I believe this may be where The Mother's people settled. If I had to guess, the mouth of the Victoria Nile is where I would start the search, it being the most likely location given the limited description we have."

The shuttle, the *Alan Shepherd*, left the hangar bay on the morning of the first day of the winter holiday season, the same day as when the old Christmas holiday had been celebrated, December 25th. They had picked out a landing spot several kilometers from the north end of the lake. The plan was to launch drones and try to locate groups of indigenous people, especially one with people with two different appearances. They knew what Eve looked like: she had fair skin, long dark hair, and beautiful green eyes. Additionally, she was tall, even for the twenty-fourth century. They hoped that her unusual appearance would make her easier to locate.

As they descended, each of the crew seemed to notice that the Sahara Desert was just not there. Instead, it contained large grassy areas and many lakes, several very large.

"The Sahara has gone through periods of wetter weather and then extreme dry weather. The end of the last wet period is still another few thousand years away," Edwin said to the quizzical faces next to him.

"This is really quite fascinating," Ed said.

"To know that this vast area the size of the United States was once greener and wetter and to see it that way are two different things. I mean, the continents look very similar, but the climate is clearly different in some areas," Ed added.

Stephanie broke in at that moment. "Folks, I am deploying the braking chutes. There is a thick atmosphere here, we are going to be bounced around for several minutes. Hang on."

The deployment of the braking chute was in fact quite a jolt, but after fully deploying, the descent became smoother as the world rushed up from below.

Their world, they all thought.

There were no cities below them, no lights, no sign of civilization; but it was the world they were familiar with, to some degree at least.

"Engaging spinner," Stephanie announced.

For the next thirty minutes, they descended further and then leveled off at about three thousand seven hundred meters. At this point, Stephanie and Dale began looking in earnest for the large clearing they had picked out from orbit.

"This might be it," Dale said as he highlighted the location on his screen so the others could see it.

"Copy that," Stephanie noted as she veered the shuttle to the south and west to put the landing zone beneath the craft.

As they descended, they noticed a large herd of buffalo grazing below.

"Look at that! Look at the size of those buffalo!" Dale said as he zoomed in on the grazing herd.

"They're huge!" Camilla said.

Ed noted, "African giant buffalo. They went extinct."

"What else will we see?" Stephanie asked.

"More than we can imagine. There were undoubtedly many other animals that went extinct that we just never found bones for," Dale added.

Camilla asked, "Will we find just *Homo sapiens*, or will there be *Neanderthals* as well?"

"We should only see *Homo sapiens* by this time," Ed answered. "Though I am preparing myself for surprises," he continued.

After they landed, Camilla sent out drones to scout three hundred sixty degrees around them.

"I have a search grid set up for the mouth of the river as it exits the lake," Dale said as he and Ed looked it over.

"It's going to be dark soon. We should get some sleep. Camilla has volunteered to be on watch first. We'll do the watch by watching the IR feeds on the sentry drones. Anything looks unusual, wake me up," Stuart said.

At 1:00 a.m. local time, the lights on the flight deck came on as Camilla announced that someone was coming toward the shuttle. Everyone got up, groggy, trying to wake themselves.

"What do we have, Camilla?" Stuart asked. "You said someone was coming toward us."

"It's a hominid. Hard to say if it is male or female. Wait, now I see a second one. I'd say this one looks female, but it's just a guess," she answered.

They all gathered around the sentry drone screen on the handheld unit that Camilla held up so they could all see it.

"They're coming right toward us," Stuart said, sounding surprised.

"They aren't slowing down or being cautious," Camilla noted.

"I don't think they are afraid of us," Stephanie said.

"Let's get outside. Be ready when they get here," Stuart said.

Stephanie and Dieter stayed inside while the others went outside to greet the visitors. Stephanie turned on the three-hundred-sixty-degree ring lights on the spinner assembly, which lit up the moonlit landscape all around the shuttle.

Stuart asked Edwin, "Any guess on the type of language they might speak?"

"The language group that people spoke in ancient Kenya was known as Cushite, though I don't speak it."

"Would the universal translator be of any use to us?" Camilla asked.

Just then, one of the two figures walking toward them raised his arm and waved then said, "Greetings, I am Seff, son of the great mother, Eve of Eden. We were expecting you. Welcome!"

"It looks like we won't need the translator," Camilla said, smiling at a stunned Stuart.

CHAPTER 92

Planning the Trek to The Mother

"SO HOW IS it that you speak English?" Stuart asked Seff.

"The Ones want us to learn the great languages of the Earth. My Mother taught me," he answered.

"But English is not even spoken yet anywhere," Stuart said.

"Yes, but you aren't even born yet, but here you are, Dr. Carroll."

"OK, you got me there. I must therefore assume that you know the reason why we are here. Do you know why we are here?" Stuart asked.

"Yes. You are no longer in your time and need help to get back. My Mother and I can help you. We have been asked to help you, to be more precise," Seff explained. "We will take you to our settlement where you will meet Eve, my Mother. It is a day-and-a-half journey, south and east of this location."

"We could fly you there with us if you would like," Stuart said.

The woman who accompanied Seff said something in a different language to him.

Their conversation went back and forth; then Seff said, "My wife, Nambi, said that it would not be appropriate for her people to see this craft. There would be questions from her people that we do not wish to answer."

Stuart thought for a moment then said, "Yes, of course, you do not want them to know of this technology or that you are familiar with higher technology. I understand. What about our clothing? It will look out of place, or rather, out of time, I suppose."

"We will make you some clothing from the great buffalo, but it will take some time to kill and tan hides. All told, our journey will extend to a week or more to accomplish this," Seff said.

Camilla looked confused then asked, "Do we stay here while you and Nambi kill some buffalo and tan the hides?"

"No, the herd we will hunt is on the way back to our settlement. You will come with us, at least the ones who need to come," Seff said as he looked around at each of them. "How many of you are there?" he asked.

"Six. There are six of us in total," Camilla answered.

"We should sleep. We can discuss who will go and who will stay in the morning," Seff said.

By nine the next morning, Seff and Nambi were awake and were waiting in the galley when Stuart and the other crew members joined them. "We've decided that myself,

Dieter, Stephanie, and Edwin will be going with you. Camilla can fly the shuttle if that becomes necessary. Dale wants to stay and take some measurements of the sun and atmosphere for Edwin, since he will be with us and won't be able to collect data he is interested in. Even though he's a psychologist, he and Edwin are interested in this rare look at the Earth before mankind became established in large numbers. But I digress. We can be ready in a few minutes. Does this sound OK to you?" Stuart asked Seff.

"Yes. We passed a large herd of buffalo on the way here. We'll need your help taking a couple of animals."

Camilla broke into the conversation and said, "Stuart, you should take a stun gun. I believe the max setting would take out a large animal like these buffalo."

"We can't risk having someone find that technology," Seff said. "Best to leave it here."

"Wait . . . this 'buffalo hunting,' is it dangerous? Those are really big animals. How can we possibly kill two of them?" Stephanie asked, clearly concerned.

"You're here. You're proof that your ancestors were can-do types of people. You're descended from the winners, the survivors. There are six of us. Odds are, we will only lose one or two of you on this hunt!" Seff said, smiling as he did.

"Well, what if it is both of you? Then we're screwed!" Stephanie said with emphasis.

"Sorry. My sense of humor may not be appropriate. We know how to hunt buffalo. There is a risk, but I've never seen anyone die from doing it. You won't die," Seff said, now switching gears from humor to a serious tone. "A side benefit from our hunt will be buffalo steaks. Nambi can make the most delicious slow-roasted fillet mignon from buffalo," Seff said, now smiling again.

Stephanie smiled also and said, "Well, I don't eat rare meat, so it has to be cooked!"

CHAPTER 93

Stone Age Hunting

FORTUNATELY, IT WAS the dry season, which meant that finding the buffalo was not an issue, Seff had explained to them. All they had to do was look for locations where the dwindling water supply still existed, and they would find buffalo and many other species of animals.

They left the shuttle at 10:30 a.m. The day warmed up rather quickly. At noon, the temperature was thirty-five centigrade (ninety-five Fahrenheit) and looked to end up above triple digits on the Fahrenheit scale. They marched single file with Nambi leading the way and Seff at the rear to make sure no one got left behind. The savannah was pocked by areas where trees grew and large areas where trees were sparse and grass predominated. At 2:00 p.m., they reached a large watering hole with thousands of animals gathered to drink, including the giant buffalo, though the herd was much smaller than Stuart expected, perhaps only fifty members. They spent the next two hours resting and taking turns practicing spear throwing with the atlatl, a prehistoric device that hunters used to greatly increase the range of their spear-throwing ability.

Seff explained how they would execute the hunt. "We will approach the herd from downwind. When we are within one hundred meters or so, we will split into three groups. The first group, Nambi and I, will hide in the tall grass. The other two groups will walk along the sides of the herd, one group on one side and the other group on the other side. As you get to within about fifty meters of the water, turn toward the herd and slowly walk toward it. Stop when you get within about thirty meters and stand where you are. Make some noise to get the herd aware of you and to scare them away from you. The water hole is very muddy—they will not move toward the water where they could get stuck in the mud. The herd should start to move away from the water. We don't want to cause a stampede, just a slow movement away from the water while you walk alongside the herd on both sides.

"At this point, Dieter, I want you and Stuart to move toward the water again. Get in behind the herd. Some of the males will stay in the rear to protect the females and young. If you have a good shot, take it. These spears and this device, the atlatl, allow you to throw a spear much farther than you could without it. Have three spears with you, throw two, keep a third in reserve in case you need it. The herd will not stampede toward the water, so you should be safe. Nambi and I will launch spears toward the moving herd, which should be coming toward us. We will throw two spears then run parallel to the moving herd and work our way back toward the water to help you out. Whatever happens, move back toward the water, which is where we all want to meet."

Half an hour later, they were in position on three sides, more or less, of the herd of fifty-some-odd giant African buffalo. Stuart didn't realize it then, but the wind direction had been from the north toward the watering hole.

Stuart thought, *How did Seff know which direction the wind would blow that January morning? What am I supposed to do if the herd stampedes? He did say that he doesn't know of anyone who was killed trying to kill a one-thousand-four-hundred-kilogram wild, horned, and ornery bovine. He would have told us what to do if he thought we needed to know it. I'll just play along. Everything will be OK, an adventure!*

And about then, the adventure began with a lot of hoof stomping.

Stuart and Edwin had begun to move toward the herd from their side, and Dieter and Stephanie were doing the same on their side. As expected, the herd took notice of them and began to move away from the waterhole and toward Nambi and Seff who were hidden in the tall grass. Stuart broke away from Edwin and walked toward the water to get in behind the herd. Dieter broke away from Stephanie and walked toward the water to also get in behind the herd with Stuart. Seff was correct that some large bulls did remain at the rear of the slowly moving herd. One very large bull turned and faced Dieter. The bull huffed and snorted at Dieter then started pawing the ground with his giant front hoofs.

"I think we should walk away from that beast!" Stuart shouted at Dieter, who had become frozen, looking at the hulking animal, who appeared enraged.

In an instant, the animal charged toward Dieter, who threw one of his spears at it without using the atlatl due to the close proximity of the beast. The spear hit the ground in front of the animal, causing it to stop momentarily. The buffalo snorted loudly and charged again. Dieter realized he couldn't hit the animal with a spear to stop it, so he began to run. The race was going to be no contest. The buffalo could run at over thirty kilometers per hour. Dieter could top out at twenty-two kilometers per hour. It was gaining quickly on him.

Dieter screamed "HELP!" as it closed in on him.

Just then, in a blur, a spear seemed to drop at supersonic speed from the sky and buried itself deep into the beast's upper shoulder. It struggled a few more steps and then collapsed in a heap at Dieter's feet. Dieter, who had fallen over, slowly got to his feet and went over to the buffalo. It was still breathing, but it was clearly mortally wounded. As he looked at the enormous creature, Nambi appeared at Dieter's side and thrust a spear into its chest, which instantly killed it. Dieter looked over at Nambi who was smiling.

"Thank you," he said in between exhausted breaths.

Nambi said something in return then turned toward Seff who came trotting up to the animal.

"That makes two. Stephanie got one also," said Seff.

"Stephanie got one?" Dieter asked somewhat incredulously.

"Yep. I threw my two spears at the herd as it moved away. One of them killed a large buffalo," she said, pointing off in the distance at another fallen animal.

"Well, that all went according to plan!" Seff said, smiling as he repeated the same thing in the native tongue to Nambi who huffed and walked toward the other fallen beast.

"I told you I had never seen anyone get killed in a buffalo hunt, though I think you came the closest yet, Dieter," Seff said as he laughed.

For his part, Dieter had now composed himself enough to manage a slight laugh; then he fainted and collapsed on the ground.

Stephanie rushed over to his side and said, "Dieter! Wake up! Are you OK?"

Dieter slowly opened his eyes, got to his hands and knees, and then lurched forward and threw up on the ground.

"I could have died! I need a few minutes," he said in disbelief. "Look at the size of that thing! And to think it could move so quickly! How on Earth did our ancestors survive?" he asked.

Edwin answered him, "We think of prehistoric people as primitive and unskilled. In reality, they had to retain a great deal of knowledge about the world around them in order to survive, and they needed highly developed skills and courage to survive in a world where they had the brains but not always the size and strength to compete with the animals in their world."

"Let's get to work skinning these animals. We have to let the hide dry in the sun for a couple of days. Then we can fashion it into clothing for each of you," Seff explained.

"Well, if I'm going to dress like a caveman, maybe I can act like one also!" Dieter said as he turned toward Stephanie and scooped her off her feet and kissed her.

Stephanie squealed in delight and said, "Dieter! We don't even have a cave to go to! What will we do?"

They all laughed at her humor as Seff pulled out a long and nasty looking knife and began to skin the animal.

CHAPTER 94

The Encampment of The Mother

THEY ENDED UP taking three days to tan the hides instead of two. The *Calypso* crew members were so exhausted after the hunt that they wanted a day to rest up.

Bruises and blisters covered the feet of each of the four visitors. General fatigue from the trek had left each moving much slower than they anticipated. Dieter did not even get off his sleeping mat for the rest of that day.

He told everyone he was considering the direction of his life and what he wanted to do with it now that he had stared down his own death and somehow avoided it.

Stuart prevailed upon Seff and Nambi to let them have a day to rest. All four of the Earthlings from the future voted in favor of resting. Seff and Nambi relented and allowed it.

As they drew near to the north end of the lake, the land became more broken.

"Why does the land seem 'broken'?" Stephanie asked.

"The elevation of the lake has been decreasing over the last several thousand years due to rifting. In essence, the land in East Africa is spreading apart and sinking," Edwin explained.

By late afternoon, they reached Eve's village. The outskirts of the village had large areas of irrigated fields of food crops and even larger grazing areas for what looked like cattle. The village itself consisted of wooden huts with many of them on stilts, presumably to protect them from seasonal flooding.

They were met by a group of women numbering around twenty ranging in age from about seven years to what seemed like about forty years of age.

"These are my sisters," Seff explained.

"There are so many of them! How on Earth did your mother have so many children? Do you have any brothers?" Stephanie asked.

Seff answered, "My mother is almost one- hundred and fifty years old, and yes, I have many brothers."

"Wow! What is her physical condition at that great age? Does she still get around easily?" Stuart asked.

"My mother does not age. She will not age past early middle age," Seff explained to a confused Stuart.

"Why is she immortal?" Edwin asked.

Seth explained, "She was the daughter of Maralee and Elotim, both of whom were Sky Lords. They are immortal, and because of that, she is immortal. My mother will explain more of these things to you if she feels you are ready to hear it."

"I want to hear whatever she wants to tell us," a clearly interested Stephanie said.

When they reached the largest building in the village, they stopped, and a woman with long dark hair descended the steps and spoke. "Welcome, my children. I am thrilled to see you *again.*"

Stuart stared for a moment at her and was made speechless by her stunning appearance. Eventually, he caught himself and responded, "It is my pleasure to make your acquaintance again, Mother Eve."

As he said this, he stepped forward and stepped up the steps, took her hand, and kissed it.

Eve smiled at Stuart for a moment then turned to all of them and said, "I know you are all weary from your journey. I have had baths drawn for each of you inside. Come in, bathe, rest, and we will talk more this evening."

"I for one am looking forward to a bath and some sleep," Stephanie said.

"May I join you?" Dieter asked.

"Of course, Dieter dear. We need to wash all that cave dust off of you!" Stephanie responded and laughed.

CHAPTER 95

An Evening of Discussions and Love

THEY WERE ALL sitting on the floor on cushions with a low table in front of them when the servants brought in the first course of the meal.

"This is very good. Is this kale and mashed potatoes?" Stuart asked Eve.

Eve looked at Stuart and smiled and answered, "Yes, we grew this in the irrigated plots you probably saw as you entered our village."

"What is the main course? Do you eat meat?" Edwin asked.

Eve answered, "The main course will be beans and buffalo steak from the buffalo you took on your way here. Nambi prepared it. Most of our meals are vegetables and fish from the lake."

Nambi smiled as she heard her name spoken. Seff was translating for her.

"At some point, we need to discuss with you the problem we encountered in coming here. I think you are familiar with the problem?" Stuart asked.

"Yes, we can talk now that dinner is finished," Eve said in answer to Stuart's question.

Eve spoke in the native tongue to one of the servants who nodded her head and said something to the other servants after which they all left. This left just the *Calypso* crew, Eve, Seff, and Nambi in the room.

"You want to know how to get back to your space-time. You are also wondering how it happened that you traveled back in time. You are here, in this time, because The Ones wanted you here. Had you proceeded to the Mars colony and Earth as you planned, you would have been destroyed by the AOW. They learned of your return trip and its timing from two spies on Mars. These two individuals are being coerced against their will to help the AOW thwart your mission. As part of your mission, you will need to help these two individuals remove the cruel implants from their brains, but this must be done carefully so that the AOW does not learn of your new arrival time. We will help you cloak your ship so that it will not be detectable by radar and telescopic devices on the Mars colony and on Earth," Eve said to the stunned *Calypso* crew members.

"Stuart, I know that you are wondering why The Ones did not prevent the AOW from learning about your return timetable. They, and I, are not allowed to prevent individuals from acting on their impulses, even their worst impulses. Each of us is free to act as we wish. It must be this way in order for all of us to grow and reach our full potential. No one is ever forced to do good or bad, ever," Eve said as she focused on Stuart's surprised expression.

There was silence in the room for a moment after Eve said this.

"Now, concerning how you came to be in this space-time location. The wrinkle that you traversed was a large wrinkle. A 'tall' wrinkle. Normally, no one should attempt to traverse such wrinkles because it can relocate travelers from their specific space-time plane. This happens because someone traversing these large wrinkles may not remain planer or in the plane of their space-time. Your ship was angled downward slightly, as it entered and traversed the space in the wrinkle. Because of that, you could not reenter your space-time.

"Instead, you entered echoes of the space-time you left. When these echoes of past moments, years, and centuries of your space-time are entered, they will put you into new locations of your space-time at times in the past. To return to your space-time at the desired time and location will require that you traverse a large wrinkle in a way that will carry you forward in your space-time. Such a wrinkle exists and is just beyond the orbit of the planet you call Neptune. The Ones have provided me the course you will follow to enter that wrinkle at a slight upward angle, which will push you into your future, two years before the AOW expects you to arrive at Mars and before they arrive to ambush you and your ship," Eve explained.

Stuart found himself mesmerized by what Eve was telling them but recovered enough to say, "Thank you, Mother Eve."

"I know you are all weary. I suggest you return to where you rested earlier. There will be beds for you there. We want you to return to your ship as soon as possible and then chart a course to the wrinkle I spoke of. The entry into the wrinkle is described in these papers that I have prepared to give to you. Stuart, would you remain behind please?" Eve asked.

"Of course," Stuart answered.

Eve remained silent for a moment staring at Stuart, then said, "Stuart, would you like to stay with me tonight?"

Stuart was caught speechless for a moment then asked, "Aren't you married?"

Eve smiled and answered, "My husband Adorn passed away many years ago. I have married others, but they have all passed on. I have been alone for a long time now. Even I can get lonely and long for a man."

"I would like that very much, Mother Eve," Stuart said as all kinds of thoughts went through his head.

"Good, let us retire to my chambers," Eve said as she grasped Stuart's hand and led him through a hallway and into her bedroom.

When they reached her bedroom, Eve slid off her robe, revealing her nude body.

Eve smiled and asked, "Do you like what you see, Stuart?"

Stuart walked over to her, sliding off the thick animal covering he was wearing, and said, "Oh yes, very much so!"

Stuart grabbed her around the waist and pulled her next to him, kissing her deeply. They both smiled and laughed then fell onto the bed and began their lovemaking.

CHAPTER 96

A Very Bumpy Ride

THE *CALYPSO* RUMBLED, something none of them had ever heard before. They were entering a heretofore unmarked wrinkle, a big one, just inside the heliosphere, far past the last gas giant, Neptune.

The crucial issue in reentering the correct space-time was the angle they passed through the no-man's-land in between the two sides of the wrinkle. Too steep of an angle, and they would enter the "blank slate" as Seff referred to it. The blank slate was the "unwritten," "yet-to-be-experienced" future space-times. They would enter nothingness. If the angle were too shallow, they would reenter in another past space-time echo. The angle had to be just right, and the calculations that determined that angle were mind-boggling.

"This is well beyond acceptable Caldonian risk limits. Are you sure that Seff was certain this will work, Stuart?" Timeye asked.

"He seemed confident, but he did say, when I asked him that this was his first attempt at calculating the 'catch angle.' We talked about this. He is a very confident young man. He was sure none of us would get killed hunting those giant buffalo, and Dieter nearly bought the farm!" Stuart answered as he dabbed at sweat dripping from his forehead into his eyes.

Captain Simmons, who had reassumed command of the *Calypso*, said to all of them, "Let's assume this was all calculated correctly. Right now, none of us know anything different. Second-guessing things now is just adding to our anxiety, and I need each of you to pay attention to your duties as we've divided them. Stuart, how much time in transit do we have remaining?"

Stuart looked down at his screen and answered, "Not much time left now, just over three minutes."

A few moments later, the rumbling stopped as abruptly as it had started. Everything was quiet, too quiet it seemed to Stuart.

"Can one of you put the forward and aft star views onto the big center screen?" Captain Simmons asked.

"Pulling it up, ma'am," Stephanie answered.

Captain Simmons exhaled audibly and asked, "What do we think, people? Do these look like the stars in the positions we expect them for the current date?"

"Captain, in anticipation of this novel wrinkle transit, I had our main AI produce a program to determine Earth dates for various key star positions," Stephanie said. "The star positions are for an Earth date of June 15, 2209, CE, which is 153 years *before* our last star date time just before we transited the wrinkle that put us back in time. This would put us in Mars orbit at about 150-some-odd years before our scheduled arrival before all this happy horseshit happened."

"How can we independently verify that date?" Captain Simmons asked.

"We could ping the Europa data link satellite. It became operational before this time. It should be working. A return ping will show a date and time," Stuart answered.

The captain turned to the corner of the room and asked, "Let's ping it. Ms. Stetson, can you do that?"

"Yes, ma'am, doing so now. It looks like it will be about a seven-hour-and-fifty-one-minute outbound time and the same in return, so about sixteen hours from now, assuming that bird is still in operation. Would you like me to check for other birds we could ping in case that one is not operating?"

"Yes, but let's wait before pinging other birds. The less signal traffic we generate the better, just in case the AOW is somehow able to listen in to what these birds are receiving and transmitting," the captain said.

* * *

"Look, Stuart, I didn't want to get everybody else worried, but why have we not heard anything yet? It's been twenty-one hours, and nothing, no return ping," Timeye asked.

Stuart listened closely then said, "Close the door, Timeye. I only want you to hear this. No one, not even the captain, should hear this, at least not yet."

"I'm listening, Stuart."

“I’m not sure we were able to precisely follow the catch angle that Seff calculated for us. Now what that means is, well, what that means is we might be in the right time frame, but we might not be in the right space-time. I know you know what that means,” Stuart said.

Timeye was silent for a moment as he thought about this then asked, “We’re in another universe, one that branched off from us somewhere back in time.”

Timeye and Stuart locked eyes and said nothing. Each of them knew what this meant.

“We may never get back to our universe,” Timeye said.

CHAPTER 97

What Can We Do Now?

CAPTAIN SIMMONS WHO was holding her baby in the seat at the front of the room asked, "Stuart, Timeye, we've received no return pings from any of the seventeen satellites we've tried. Nothing from any of them. So first, why is that? Second, what can we do now to determine *when* we are?"

Timeye and Stuart locked eyes. Then Timeye answered, "Sandra, Stuart and I have talked about this. There is no easy way to say this, so here goes. We believe we are in a completely different universe, a universe that branched off from ours some time ago."

"OK, does Earth even exist in this universe? How about the Mars colony? Europa? Can we ascertain *when* this branching occurred?" Sandra asked.

Stuart answered, "I've looked at Earth through our optically enhanced light-gathering telescope. This is like the ones used to look at exoplanets to determine the composition of their atmospheres. Earth *does* exist. In fact, there are zero radioactive isotopes, beyond the naturally occurring ones, in its atmosphere."

Stephanie perked up at hearing this and asked, "Wait, what? Why would the atmosphere be 'clean'? Earth has had significant radioisotopes in the atmosphere for hundreds of years, from the nuclear wars and the spreading of isotopes from the destruction of nuclear power plants during those wars."

"It is my guess," Stuart said then paused, "that those wars never happened on this Earth in this universe. Otherwise, we would see those isotopes."

Anna Stetson, normally always cheerful but not so now, asked, "What else do we know? What about television and radio broadcasts? They should tell us something, right?"

Dieter stood to answer. "I've been looking for those. Stuart asked me to do this. There are no broadcasts after 2033 CE. The ones I have listened to are, well, puzzling and may indicate when this branching occurred. Al Gore won the 2000 election, because Florida was recounted at the direction of the US Supreme Court. I queried the history AI, and in *our* history, the US Supreme Court stopped the Florida recount resulting in George Bush's victory."

Sandra's eyebrows were furrowed, deep in thought. Then she asked, "What about the War on Terror? Was there one? Do you know anything about the Traitor, climate change? Did any of that happen?"

Dieter answered, "I find references to a briefing given to President Gore by George Tenet, the CIA chief in August 2001. President Gore was briefed on the plot to fly airplanes into buildings by Al-Queda. He enacted emergency security measures for the aviation industry, which were adopted worldwide. Additionally, the Al-Qaeda terrorists who were training at US aviation schools were arrested and interrogated. One of them confessed to the details of the plan and of the leaders of Al-Qaeda. A SEAL Team Six special operation in Afghanistan resulted in the arrest of Osama Bin Laden. He was tried in a US court and imprisoned in the supermax incarceration facility in Florence, Colorado. He died there in 2027.

"In our timeline, President George Bush was also briefed on the Al-Qaeda plot by George Tenet, but infamously told George Tenet, 'Well, George, you've covered your butt now, and I'm now covering mine,' but did nothing to prevent these tragedies. We all know what happened next. Incidentally, I find no references to the Great Recession. There was still a pandemic in 2020, but President Hillary Clinton deftly handled the crisis with only thirty thousand American deaths, not the 1.19 million deaths that occurred in our timeline due to the incompetence of the Traitor. One more thing, research into EPC accelerated in the 2020s, resulting in working prototypes in 2026. I have asked myself if this could be the reason we see no TV or radio broadcasts after 2033. I believe it may be so, or perhaps these signals were encrypted by legislative mandate around that time."

Sandra and everyone were silent for a moment after Dieter finished. Then she asked, without looking up at the others, "So there may have been no need to build the *Calypso* and save the species. We have things figured out in this universe. If we go back to Earth, what else will we find? Will any of us exist?"

Timeye sensed that his wife was deeply disturbed and answered, "Sandra, none of us know what else we would find out. But perhaps . . ." He paused as he stood up and walked to her side and knelt next to her.

"Perhaps they will know how to get us back to our universe. Obviously, they are much more advanced technologically than we are."

Sandra listened, hesitated momentarily, then said, "Stephanie, Ms. Stetson, set a course for Earth."

CHAPTER 98

Planning the Approach

"WE CAN'T JUST move the *Calypso* into Earth orbit—we all know that. The risk to the planet of having a rock this big in orbit around the Earth will simply not be allowed by Earth authorities. We know nothing about the level of technology of this Earth, but it is apparently more advanced than where our Earth was at the same time. So we need to find an orbit that passes somewhat close to Earth, close enough for our shuttles to reach Earth and land there," Sandra stated.

"I think we need to assume that they can detect our presence out to probably one AU. Any closer and they may take action against us. Our shuttles can travel a Mars-to-Earth type of distance, with a low consumption mode of travel. We are at 0.04 percent of light speed. The shuttle will leave with that velocity, so they do not need to accelerate much more to reach Earth in a decent time frame," Stuart added.

"That would be about a seventeen-day trip at closest approach if we enter the Mars-to-sun orbit Stephanie showed us," Timeye said.

"What about the return trip? Will we need to decelerate to allow the shuttle to catch up with us?" Sandra asked.

Stephanie answered, "The shuttle will need to have a full tank of hydrazine fuel to boost their velocity and escape the Earth's gravity. They'll need help from the people on Earth for that, no way around it. It's a risk we have to take, in my opinion.

"To catch up to the *Calypso*, the shuttle will need to act as if it were going to Mars. The Hohmann transfer orbit has one window every twenty-six months. Luck is on our side. Upon Earth arrival, we will have eight months to wait, not twenty-six months.

"Our approach to Mars and our pull away from Mars would be our fastest point in terms of velocity. As we pull away from Mars, we will start to slow down. The shuttle will also be accelerating as it rounds Mars. At that point, the *Calypso* will need to decelerate to allow the shuttle to catch up. This will change the *Calypso*'s orbit. We will need to use a set of solid booster rockets to regain that velocity after the shuttle returns to put us back into the same orbit. It is doable. The return trip to Mars and rendezvous with the *Calypso* will be nine months for the shuttle."

Sandra was silent, staring at the star viewer screen in the conference room, then asked, "OK, who wants to go?"

"I think Dieter and I should go, as well as Stuart," Stephanie said.

"I want to go as well," Carlo Bertucci said. "I'm a historian. It may be useful to know what happened on Earth and what is the deviation from our timeline."

The captain stood and said, "Sounds good. When does the shuttle need to launch?"

"Thirty-six days from now. We'll need to get the shuttle ready. We want a full tank of liquid fuel, two sets of descent parachutes, food, and other supplies. Also, we need to decide where to land and how we will approach the people there about our problem. I mean, one thing we haven't talked about yet is whether they will even be able to help us or whether they will want to help us," Stuart said.

"Yes, that is really the crux of the matter here. We should try to think this through and get back together to discuss it. Also, I want to go as well," Timeye said.

Sandra's face changed instantly upon hearing that Timeye wanted to go.

"I may need you here, Timeye," she said.

"Stuart's brother, Carl, will be able to help you in the ways I have," Timeye countered.

"Let me think about that, Timeye," Sandra answered.

CHAPTER 99

Prelaunch Briefing

THIRTY-SIX DAYS CAME and went fast, too fast. None of us know what this Earth will be like. Will we even find the help that we need to get back? Back to our universe. The fucked-up one where we ruined our Earth. Where we fought needless wars? One where we fought TWO *nuclear wars and destroyed the entire ecosystem? Why were these people smarter than us?* Sandra wondered as she lay in bed next to Timeye.

Sandra got up, showered, dressed, ate, and then headed down to the officer's conference room. Today's briefing was the last chance to go over the plan for the new Earth expedition.

* * *

As the captain walked into the conference room, everyone stood. "OK, people, let's run through our strategy one more time. Stuart, run us through this."

Stuart looked down at his tablet and then began. "We have months on the planet. We will begin by landing in the Glacier National Park at this location." Stuart pointed to the holographic map and long-range reconnaissance pictures they had obtained from their telescopes. "It looks pretty much like what we see on our Earth pictures of the park. We've identified this location, which is about fourteen miles from the Going-to-the-Sun Road. There are several locations for us to pick from for a landing area, near where I am pointing to. Once we land, we will monitor any communications we can and then decide where we will go next. We anticipate that we may go to either Missoula, Spokane, or even Salt Lake City in our hovercar. Stephanie will stay behind with the shuttle.

"Our cover story will be that we are college professors on a trip to Yellowstone and the Grand Canyon. We need to get some local currency, whatever form that is in. We're bringing several items that we think we can sell. Once we reach a larger metropolitan area, we will get access to whatever passes for their internet or Webcon for any articles that may relate to extra universe travel.

"At some point, we may have to travel to wherever we need to go to contact individuals who have expertise in this subject. We plan to keep our true identities and the shuttle location a secret unless we feel it is safe to reveal it. We'll use our real first names, but we will change our last names. The current date, as far as we can tell, is 2209, a hundred-and-thirty-some-odd years before we left our Earth.

"So much of what we will be doing will be ad-lib. We will take our time to learn as much as possible about the culture and history of this Earth so that we can fit in better. Oh yes, one more thing. Each of us will be carrying stun handguns, just in case."

The meeting continued on for another hour. Intercepted communications were very sparse from this Earth, leaving many questions unanswered.

CHAPTER 100

Landing

THE SHUTTLE HAD completed its seventeen-day journey to Earth. In front of them, the world they all knew loomed large. So far, no one had tried to communicate with them.

"I am a tad bit surprised no one has tried to communicate with us. We have detected numerous satellites, but none of them seem to have alerted Earth authorities to our presence. I'm suspicious," Stuart said.

"Well, if they know about us or not, we are going to land in the location we have planned," Stephanie said.

Stuart, who was sitting in the copilot's seat, frowned and then refocused on his job as a copilot.

Dieter, Carlo, and Timeye were seated in the jump seats on the top level. Everyone could see the large viewing window. The rotors were stored in the up position, giving the shuttle the shape of a cone as it descended through the atmosphere.

"Deploying braking chute," Stephanie said as everyone was jolted by the deployment of the parachute.

A few moments later, the parachute slowed them down enough and they were low enough that the rotors could be lowered and engaged, putting the shuttle in spinner mode.

"Eighteen thousand meters, rotors lowered, spinner engaged!" Stephanie said as Stuart was now visibly more relaxed in the copilot seat.

"Little nervous, Stuart?" Stephanie smiled as she asked this.

"I always feel better the closer we get to the ground and a safe landing," Stuart answered.

Stephanie turned to those in the jump seats and said, "Well, we aren't safely on the ground yet. The computer will take us to the landing coordinates, but I will be taking over manual control when we get low enough as the observations we made from space are just a best guess at a landing site.

"Seven thousand meters altitude. I have the holographic cursor on the large viewing window showing our landing site. If anybody notices anything I should know, please speak up."

"I see a clearing in the forest. It looks level, and I see no human activity. It looks good to me," Timeye said from behind Stephanie and Stuart.

"I concur. The Road-to-the-Sun is a good twenty-two kilometers' distant; I can see it. I think we will have a good chance of being unobserved there," Dieter said.

The setting sun was now sliding below the horizon. It would soon be dark.

“We will set down there. We only have a few minutes of light left, we can’t take any more time looking around,” Stephanie said.

The shuttle lowered to the ground, making hardly a noise as it did so. They were on Earth again though it was not their Earth or their time.

CHAPTER 101

Hitching a Ride

DIETER, CARLO, STUART, and Timeye crouched in the bushes, keeping themselves hidden from any traffic on the road.

"I've seen nothing but hovercars for the last thirty minutes. No ground vehicles at all. I guess this old road is just a flight corridor for hovercraft, at least inside the park area," Timeye wondered.

"Yeah, we would stand out like a sore thumb with our hovercraft. It is much bigger. Maybe we could hitch a ride?" Carlo asked.

"What does 'hitch a ride' mean?" Timeye asked everyone.

"It means we try to get someone's attention and then ask them if they can take us where we want to go," Stuart answered.

"Carlo, if you go out on the road and wave at passing hovercraft, maybe they will stop. You're a handsome Italian guy, maybe the ladies will stop for you," Dieter said with a smile.

"OK, here goes nothing," Carlo said, exhaling deeply and audibly.

The road was empty for nearly fifteen minutes before a hover car came by. Carlo waved and smiled as the others watched in a small group ten feet away from him. The hovercar slowed down and stopped.

A mature woman, probably forty years old, lowered a window and asked, "Go you where, fellow traveler?"

Carlo smiled at the curious twist in the language.

"Friends and I go where you go. Go you where, lady beautiful?"

"Speak you strange, fellow traveler! Speaker of old English, eh?"

"Yes, I am from Italy." Carlo finally gave up on trying to fake the odd dialect.

The woman smiled and said, "I can speak old English also. Are you Fundamentalists?"

Carlo thought for a moment. *Fundamentalists? Christian Fundamentalists! Yes, I remember some things about these odd Christians from the twentieth and twenty-first centuries. I could say we are Fundamentalist Mormons. That might get us to Salt Lake City.*

Carlo decided to go with his hunch and said, "Yes, we are from a colony of Fundamentalists. We are college professors. We would really like to go to Salt Lake City. Perhaps you are going that direction?"

The woman frowned, paused, then said, "I'm going to Pocatello. I could get you that far. There are so few people left in Salt Lake City. The toxins from the old lake drove most people away, though many have returned recently, I have read. Get in, I can take all of you though it may be a bit crowded."

After they all got into the car, the woman introduced herself. "I am Maria Welch. I am a professor of linguistics at Idaho State University. Also, you are obviously not who you say you are. Could you tell me who you really are and what you are doing? I may be able to help you if you are honest with me."

The *Calypso* crew members stared at one another, and for a few seconds, no one spoke. Then Stuart said, "I am Stuart Carroll. I am a physicist on the Free Bloc Ship *Calypso.* We landed in our shuttle craft a few kilometers from here. We are here because we think we can obtain help in returning to our home."

"Stuart, where is your home?"

"Maria, we are from the planet Earth. Though not from *this Earth.* Also, we are from the year 2365 CE. We traveled to a star called Cervantes and have settled on a planet we named Terra Nova. We were trying to return to rescue survivors on the Mars colony and on Earth when we traversed a wrinkle in space-time and entered a new universe. *Your universe.*"

Now Maria was silent for a moment. The hovercar drove on in autopilot mode. Maria turned and looked at Stuart in the front passenger seat and then at those in the backseat.

"As outlandish as your story sounds, I believe you. Why, you might ask? Well, why would you tell me such a strange story if it were only a lie? It must be the truth.

Now I must ask, what can I do to help you? Please be as specific as possible."

Stuart cleared his throat then said, "*Your Earth* appears to be at least as advanced as ours was even though we are from the future. The only thing we could think to try was to come here and see if there were theories and technology that may allow interuniverse travel. Do you know of anything in your knowledge base that may help us?"

Maria thought for a moment then said, "I am a linguistics professor, not a physics professor. However, I have a friend in the physics department, Dr. Levvi Markanen. He has written on this subject. He may be of some assistance. He is young, brilliant, and very tall, over 213 centimeters. He is a visiting professor from Finland. He should be able to help you."

Stuart and Timeye exchanged a glance at each other; then Timeye said, "That sounds like a good place to start. Thank you."

"Timeye, did I pronounce your name correctly? You do not have the same accent as Stuart. Where are you from?" Maria asked.

"Yes, you pronounced my name correctly. I am from a planet called Caldonia. It is many light-years from here. The *Calypso* crew rescued me and some colleagues while they were en route to Terra Nova."

Maria's eyebrows jumped up and she asked, "You're not from Earth! Are you an alien?"

"Well yes, I am actually," Timeye answered.

"How did you learn to speak English? Do you speak other Earth languages as well?"

"I learned by listening to your broadcasts, and yes, I can speak several others, including Spanish and Chinese."

"Well, English is spoken everywhere here now. After the Great Unification, everyone eventually spoke English. That's been over a hundred and twenty years ago now. I take it on your Earth there must have been no Great Unification?" Maria asked with a shocked look on her face.

"I guess you're going to have to explain that one to me," Stuart said.

"Fascinating! You can all stay at my place. I have so many questions!" Maria said excitedly.

"Yes, and we have many questions as well. I am a historian, so it will be interesting to learn how things changed from our point of divergence," Carlo said.

Maria looked confused. "Point of divergence? What is that?"

"Sounds like we all have so many questions for each other," Timeye said as he smiled.

PART FOUR

Becoming The Mother

CHAPTER 102

Hunting

HATCHER WAS DRESSED in the garb of the Akeye clan—leather chaps with a large woolen coat and laced leather boots. Skrill, who had taken the Earth name Joshua, led the three hunters as they stalked a herd of large woolly mammals. The mammals looked like bison that once roamed over large ranges across three continents of Earth.

"Take care, my friends. Sage cats roam here also in search of, what you call it? Buffalo?" Joshua asked.

"Yes, buffalo, but you don't need to call it what we called it," Hatcher answered in a hushed tone.

"People of Akeye clan, we adopt your language and ways," Thomas said.

Thomas was a younger brother of Joshua, who had also chosen an Earth name.

"Thomas, who is the mate of your sister? She has a child, but I have not met her man," Hatcher asked.

"Her man was killed by sage cat on a hunt. She has no one," Thomas answered.

"It is good that you all help take care of her and her baby. Do members of your clan choose mates from other clans or just within your clan?" Hatcher asked.

"We have no free males in our clan. She will remain alone unless one of your people takes her. We think of your people as part of our clan. She will have a hard life unless one of your people chooses her. Do you wish to have her?" Joshua asked this time.

"I already have a woman. Her name is Terra," Hatcher answered.

"Yes, but your woman is a great mother and leader. She can give more wives, no?" Joshua asked.

"It is not our custom to have more than one husband or wife," Hatcher said, confused with the direction of the conversation.

"Among our clans, the great Mothers can give their men more wives. More wives help the great Mother do her work in leading the clan. Terra should give you more women. She should lead and let your other women care for the children. Terra is great with child, now is a good time to have her give you another woman," Thomas chipped in this time.

"I will ask Tenaw. I am sure she will let you take her as a wife. Then you ask Terra if you can take a new woman to help her with your babies," Joshua said.

Hatcher was at a loss for words. The clan still did not understand that Terra was technically not the leader of the Earth people. Hatcher had explained that she was an adviser to President Davis, but the Terrans easily observed that Terra was loved and respected by the Earth people as the de facto leader despite the title given to President Davis.

At that moment, they all froze as they heard a slight rustling in the brush near them.

"Sage cat," Joshua whispered.

Hatcher pulled out his stun rifle, ready to fire at whatever came from the brush. In a blinding flash, a shape leaped from the bushes and knocked Hatcher over, sending his stun gun flying out of his hands. The sage cat was enormous, bigger than a tiger. Hatcher managed to get his knife out and repeatedly stabbed at the animal as it bit and clawed him. He was in a fight for his life, and he knew he was losing. Joshua and Thomas shouted at the enormous cat, which succeeded in drawing the animal off Hatcher's badly injured body. He knew that the cat would finish all of them off.

Hatcher crawled, bleeding and with one arm dragging limply behind him, to the stun gun. He slid the stun setting to MAX, pointed at the cat, and fired. The pulse of static electricity engulfed the cat and the two Terrans just behind it. In an instant, the cat fell limp on the ground, having taken the majority of the blast. Hatcher crawled to his hunting companions and saw that they were unconscious. He knew the cat would regain consciousness in only a moment and would kill them all. Hatcher crawled to the cat, raised his dagger, and plunged it into the animal's eyeball, killing it.

Hatcher reached for his limp arm, pulled it close to his mouth, and spoke into his colony comm device. "Need immediate medical assistance. Life-threatening injuries, my location, come . . ." He fainted from blood loss without finishing the call.

CHAPTER 103

The Golden Moments

WITHIN MINUTES, TWO EMTs arrived via hover vehicle at the location of the injured hunters.

"Clear, shock at two hundred forty." The EMT discharged the voltage directly into Hatcher's lifeless body. "Still nothing. Try three hundred this time." Hatcher's body jumped again from the electrical discharge.

"I've got a heartbeat. We need to get him back *stat*. He's lost an enormous amount of blood. I'll load him into the hovercraft. Eli, you stay with the Terrans. Radio in for additional assistance," Candy, the other EMT, said. The hovercraft rose up and then darted off toward Benston.

"Benston medical, code blue, code blue. It's Hatcher. He's bleeding out and lost a pulse. We reestablished a rhythm, but it is unsteady. Has Eli requested more assistance at his location? He has two injured Terrans left at that location, over."

"Copy that, medevac assistance is en route. What is your ETA, over?" the medical center dispatch asked.

"Just over a minute. Wait, I've lost him again! Be ready when I arrive. I'm not stopping to get his rhythm back.

Be ready upon my arrival!" Candy shouted, adrenaline pumping through her.

The hovercraft roared over the colony walls, barely clearing them, and skidded to a stop at the side of a waiting medical team. Hatcher was taken off and put on a gurney while a nurse jumped onto the gurney and began a furious attempt at CPR as the team pushed the stretcher into the medical hut. Just then, Terra ran into the trauma bay, asking for Hatcher.

"We've got him. His heart stopped, and he's had enormous blood loss. Terra, you're going to have to wait outside of the trauma bay," a nurse told her.

Terra began sobbing as she tried to push toward the gurney that Hatcher was on.

Simi reached for Terra and told her, "Go sit in the waiting room. I will go help him."

Simi ran into the trauma bay as the nurse administering CPR jumped off the gurney and Hatcher was transferred to the examination table. "Still no pulse. I have no pulse. Shock him again at four hundred," the lead trauma surgeon, Dr. Lynn Sampson, said.

The shock was administered, but no heartbeat was detected.

"Want to call it, Doc? I've got 12:02," a nurse said.

As he said this, Simi reached out her hand and placed it on Hatcher's chest. The monitor began beeping, and Hatcher's chest started to weakly rise and fall as he breathed for the first time in several minutes. The doctors and nurses looked at Hatcher with relief evident in their eyes. In the fuss over Hatcher, no one noticed for a moment as Simi dropped to the floor.

"Get her onto a table. Dr. Sampson, she has a very weak pulse. Should I start an IV?" the lead nurse, Beth, asked.

"Raise her legs up. Let's monitor her heart and breathing. Get an IV started. What exactly happened to her?" Dr. Sampson asked.

"She came into the room, and I saw her put her hand on Hatcher's chest. That was right when he started breathing on his own. I'm not sure why that would have caused her to faint, Doctor," Beth said.

Gunner, who had quietly slipped into the room unnoticed, said, "She's a Caldonian healer. When she focuses and heals someone, she shares her energy, which can cause her bodily distress, dependent on the seriousness of the injury of the person she shares her energy with. Will she be OK?"

"I really don't know anything about what you just told me. I can't answer that, Gunner," Dr. Sampson said.

For the next hour, the hospital staff carefully monitored Simi. As time passed, her condition stabilized and improved. An hour later, she opened her eyes and asked how Hatcher was.

"We're continuing to monitor his condition. We are transfusing him with two units of blood. He's breathing and has a heartbeat," Beth answered.

Gunner and Simi looked at each other with relief. Simi said, "I rest now." She closed her eyes.

CHAPTER 104

Recovery

"IF THE *CALYPSO* were still here, we could have transported you up to the ship and saved your arm. Unfortunately, we didn't have their facilities available to us. I'm so sorry for the loss of your arm, Hatcher. We are going to try a stem cell regeneration of your arm at the shoulder nerve node. We will be implanting a micro voltage stimulator in the shoulder nerve bundle between there and where your brain map shows that your left arm is controlled. The constant stimulation will make the stem cells know how to grow nerves back down your arm and into your hand. The muscle, bone, skin, and other cells will come after that and will be done by running a stimulator from your right arm over to your left arm and reversing the polarity to account for the side difference. There is a decent chance this will work, though it will take twelve to eighteen months before we'll know if it is successful," Dr. Sampson said in her non-emotional tone.

"We're just glad he is alive, Dr. Sampson," Terra said with more than a few tears in her eyes.

For his part, Hatcher sat motionless, looking down at his feet.

When the doctor left, Hatcher lifted his head up and spoke to Terra. "What if you asked Eve? Their technology is far superior to ours. Maybe they could produce a better result."

"I will ask her though she has told me before that we would receive help with establishing the Terrans as an intelligent, virtuous species, but concerning everything else, we would be acting on our own. You know why that is. We are still evolving toward our destiny, and we must work it out on our own. But I will ask, my love," Terra said with concern for Hatcher bleeding through her words.

Two weeks later, at a follow-up visit, Dr. Sampson began the brain mapping. The session was tedious, lasting the entire day.

"Hatcher, let's do that again. I want you to imagine making a fist with your *left* hand. Squeeze tightly, now release, and straighten out your fingers. Imagine the feeling of relaxing your fingers as they straighten out. Good, I think I've got that located. I think this is enough for today. We'll repeat this two more times over the next two weeks to double and triple check the brain locations for all the arm movements and sensations. I'm more confident about this, Hatcher. I'm not a specialist in limb regeneration, but I do have the specialist AI to help me out, and I'm surprised at how good it is. You don't need to feel like a guinea pig. This is how specialists were trained on Earth. Nothing like on-the-job training and experience."

"I appreciate all you are doing for me, Doc. I don't feel like this is fair to other patients of yours. This takes so much of your time," Hatcher said.

"Well, this is much more interesting than setting broken bones and removing appendices," Dr. Sampson said. "Besides, we need to establish this capability in our colony. There will be other occasions when this will be necessary."

Over the next few weeks, Dr. Sampson repeated the brain mapping, confirming locations to insert electrodes into Hatcher's brain and running them to the nerve node in the shoulder. The day came six weeks after the incident for the implantation of the brain micro leads. The surgery took thirty-six hours and required every other doctor and twelve nurses to perform.

A week later, Hatcher was released from the hospital and instructed to begin a regimen of daily movements to stimulate the proper growth of nerves down into where his new arm would be. A tubelike device with cell-nutrient solution inside was attached to Hatcher's shoulder and secured to his side. This allowed the nerves to grow to an approximately correct position. So far, everything was going according to Dr. Sampson's plan. The road ahead was still very long, and even Dr. Sampson tempered her original optimism, given the ever-growing complexity of the path toward regeneration of Hatcher's new left arm.

CHAPTER 105

Domestic Help

"HATCHER, CAN YOU get me another of those small pillows? My back is so sore," a very pregnant Terra asked in her sweet can't-say-no voice.

"Sure, let me get one for you," Hatcher answered.

"I get it. You stay here, keep resting," Tenaw answered. Tenaw had shown up at their doorstep a week ago with her own eighteen-month-old toddler in tow. Hatcher was caught completely by surprise as he, of course, had never told Joshua and Thomas that he had agreed to anything about taking a second woman. He certainly had never asked Terra about it, yet she seemed to have a serene and unruffled viewpoint on it once he got Terra alone and explained to her what it was all about. Terra had simply said that it made sense, that Tenaw seemed like a very nice and helpful young woman and that Hatcher could use the help and companionship. Hatcher was completely caught off guard by Terra's attitude, and he repeatedly tried to tell her that he simply did not want another woman, to no avail. Tenaw, who had said she wanted to be called "Taylor," had stayed. Each of the last seven nights, she had put her toddler to bed and climbed in bed with him and Terra. Hatcher did not know what to do as Taylor would always climb into bed next to him, nude and smiling. Hatcher had made excuses each night to forestall any thought of sex with her.

Finally, on the seventh night, Terra and Taylor both got into bed, nude, together. What happened next was a complete surprise to Hatcher.

Both women threw the covers off the bed, exposing Hatcher, who was in his boxer shorts. The women stood and started caressing each other and kissing. Quiet muffled moans of pleasure came from each. Terra lay down on the bed, and Taylor continued caressing her. With each kiss, Terra's pleasured moans became louder. After a few minutes, Terra's moans reached a climax as her entire body quivered. Hatcher watched as Terra's shaking finally stopped. Terra then grabbed a pillow and a blanket and was asleep within a minute.

Taylor came over to Hatcher, smiling and giggling. She pulled off his boxer shorts. At first, Hatcher tried to pull away, but it was difficult for him to move since he had a cumbersome prosthetic tube attached to his left side. Taylor smiled and gently grabbed Hatcher's right hand and inserted his fingers into her. Taylor began moving her hips back and forth, which caused Hatcher's fingers to go in and out of her. A couple of moments later, Taylor arched her back and cried out in a pleasurable moan. She then straddled Hatcher and began grinding forward and backward on top of him until he exploded in pleasure. Hatcher turned toward Terra and cuddled his sleeping wife. In a moment, he was also asleep with Taylor hugging his back and her arms around him.

Hatcher woke in the morning with neither woman in bed with him. He stood up, groggy, and realized he was naked. He froze in panic as he remembered the events of last night.

Shit! What if Terra saw what had happened after she fell asleep? Hatcher thought. *There is no way she could have slept through Taylor's moans of pleasure. She knows! She's left me! I need to find her and try to make this right somehow!*

He quickly dressed, which, for his shirt, was impossible with the prosthetic tube. He ran into the kitchen to find Taylor and Terra smiling and talking quietly while fixing scrambled eggs. Taylor's baby was in a booster seat, fiddling with a pile of scrambled eggs, dropping most of them onto the floor. Both women turned to face Hatcher, who must have been red as a beet, as they both began giggling.

"What's going on?" Hatcher asked, exasperated.

They both answered in unison, smiling.

"How did you sleep, *honey*?"

Terra smiled even bigger as she emphasized the word *honey*.

"Um, . . . ah, fine, good," Hatcher said, stumbling over his words, now turning even more red.

"Have some eggs, *sweetheart*," Taylor said in her Terran accent, sexily drawing out the word.

Hatcher stared at both women, confused. The deer-in-the-headlights look on his face must have been obvious as they both burst out laughing at him.

CHAPTER 106

Growing Pains

"TERRA, TAYLOR, WOULD either or both of you like to take a walk to the beach?" Hatcher asked, bored with being trapped inside.

"I put Zachary to sleep. Maybe in the morning?" Taylor answered.

Terra stood with the palm of her left hand on her lower back and her belly protruding out and said, "I'll go with you. Walking seems to help my back feel better."

"Do you ever think of your grandparents and aunt?" Hatcher asked.

"I do, but it is the same longing that I know everyone else feels. I suppose that makes it easier. Misery, it is said, likes company," Terra explained.

The path to the beach was well-worn by now as many of the colonists would walk to the beach to take a break and enjoy the sunrise or sunset.

They walked in silence the rest of the way, each in their own thoughts. Hatcher still wondered if his arm would ever grow into a real replacement for the one he lost.

I know I should feel fortunate to live in the twenty-fourth century. In prior centuries, I would have just been a cripple. This skinny, short arm—will it ever look like the one I lost? Hatcher thought.

Terra, sensing Hatcher's thoughts, said, "I know you are worried about your arm, but I believe you will eventually have a strong new arm, Hatcher."

"I suppose you could read my thoughts, huh, honey?" Hatcher said.

"Well yes, but I think I could have read your thoughts even without this gift. You're not that hard to read, sweetheart," Terra said as she playfully bumped Hatcher off the path with her superior pregnant body mass.

They reached the beach just as the setting sun was dipping below the western hills. The long, low rays of the sun reached out into the water, bringing clarity that only comes when a day is nearly done and the sun shines for a few last minutes on everything.

They sat on the beach, watching the gentle waves roll up close to where they sat. The sound of the waves became hypnotic, causing both to feel drowsy.

"I think we should go back home now. I don't have the gift of seeing the path in total darkness. How 'bout you? Can you see in the dark, Hatcher?"

"Not really. I was hoping you could." Hatcher snorted and laughed.

The next morning, Hatcher got up, leaving Terra sleeping.

He whispered to Taylor, who lay next to him, "Let's go watch the sunrise. Zachary will stay sleeping. He was up late last night."

Taylor smiled, got up, and stretched out her long, lean body, all the while smiling at Hatcher, tempting him with her nude body.

"Maybe we will have some time at the beach for fun," Hatcher said, now smiling back at her.

Taylor put on a robe and sandals, and they walked out the door, holding hands.

When they reached the water, the sun was rising far out on the eastern horizon, its golden rays reaching toward the land, slowly lighting up everything. The freshness of a new day was exhilarating. Hatcher pulled Taylor close to him and pulled off her robe. They lay on the sand and made love.

A few minutes after they finished, both fell asleep. They were awakened by children running on the sand. Taylor hurriedly pulled her robe on while they both laughed and made a hasty retreat to their house.

* * *

Terra's pregnancy did not go full term. At thirty-seven weeks, she went into labor. After twelve hours of labor, she delivered twins, who were named Constance and William. Almost immediately, their parents started referring to them as Connie and Willy.

The cries of babies could be heard when walking down any street of the colony. Throughout the colony, more and more were being born. The colony was growing by about four or five babies a month. Everyone seemed to be getting in on this action, which kept the medical staff well occupied and made sleep-deprived zombies out of many a colonist. Most of the couples were first-time parents. Children were the future of the colony, and the community prioritized family life by creating support groups that provided day care for the new parents. This allowed the parents to continue at least part-time in their occupations and work duties.

Since the human genome had been modified in the Free Bloc States over a century earlier to minimize genes that caused greed and selfishness, it had been much easier for the FBS and now the colony government to pass legislation that helped families nurture and raise children. The twentieth and twenty-first centuries were viewed as the last great hurrah of the greed-and-selfishness-based political ideology. The last two centuries had stabilized societal structures, and family life burdens were eased and shared by all in society. This resulted in far more successful parenting outcomes, which lowered crime rates, almost eliminating juvenile crimes entirely.

CHAPTER 107

Expeditions

AS YEAR THREE started, a meeting of The Mother's council was held to determine the status of establishing native civilizations.

"Terra, what, if any, additional input have you received from Eve beyond what was in your synopsis?" Dr. Larsen asked.

"The synopsis is complete by itself. Eve has reviewed it and approved it. I don't have anything else to add," Terra answered.

"I think we need to take a survey of the hominid groups on this continent," Dale LaPlant said. "So far, everything we know is based on the dozen or so clans that we have located here near Benston. None of these clans have any knowledge of clans interior to this continent. I know the two primary objectives for the first decade are to build relationships with the local clans and establish common beliefs and traditions between our colony and the clans. There may be traditions and beliefs among clans farther distant from our locations that would be useful to integrate into this initial civilizational node.

"When our civilization here eventually bumps into more distant clans, the outcomes will be enhanced if they discover some shared beliefs. I propose we can make three two-month duration expeditions to the interior of this continent. I put together a plan that I posted to our under-consideration directory. I know many of you have looked at it. The goal would be to get a read on the beliefs of these other groups and check for beliefs we can integrate into our clans here. It could pay dividends over the next couple of hundred years."

"We could put it to a vote. I think there is some sound logic there," Terra said.

"We'll just do a voice vote. All those in favor of the plan that Dale presented indicate with an aye. Those opposed, any nays? No nays then. Dale, get with security and set this thing up. Let us know the roster of people who will be involved and what the detailed plan is in terms of when we would go and so forth," Dr. Larsen said as he gaveled the monthly meeting to a close.

* * *

"How did physical therapy go, honey?" Terra asked Hatcher.

"Good, my arm is about 75 percent of where my old arm was. They keep telling me it will never be 100 percent, and that 90 percent is the best case. I guess I can live with that."

"Well, it looks pretty good. I guess I didn't know what to expect from an aesthetic viewpoint, but it certainly looks real."

"Well, it is real, just new." Hatcher said.

"I saw you talking with Gunner and Dale. Did they ask you to help with something?" Terra asked.

"Yes, and I don't think you are going to like it. They want to me go on this first recon mission. I'd be gone two months—"

"I know about it. Are you sure you are ready to go out on these kinds of things, sweetie?" Terra asked.

"Yes. I've got to start doing things again, things I am good at. Otherwise, well, I don't know how I can feel good about myself," Hatcher said.

"OK, but I'm going to go with you. I don't want you to go, but if I go with you, I'll feel like I have some control over things. I just don't want to lose you or see you get hurt again," Terra said, pushing back tears.

"Hon, you have the kids to watch," Hatcher said. "They need you."

"Taylor can watch the kids. Simi and Taylor are best friends. We can trust them," Terra said defiantly.

"I know, I know we can, but what will everyone here think if The Mother's representative leaves to go on a mission like this?" Hatcher asked, summoning his best argument against her going.

"Well, they will have to get used to the idea that my job is to make an intelligent and virtuous species on this planet, and that is what I would be doing. I have Eve's promise that I will not die or age. Do you have that, Hatcher? Do you?" Terra sobbed.

"No, but I'll be fine," Hatcher said. "OK, I can see you are going to hold on to this. I agree, but now I'm going to feel like I must protect you."

"Don't worry about me, Hatcher. I'm a big girl. I can take care of myself!" Terra said indignantly.

Hatcher shut up, knowing that she could indeed take care of herself.

The Western Hemisphere of Terra Nova

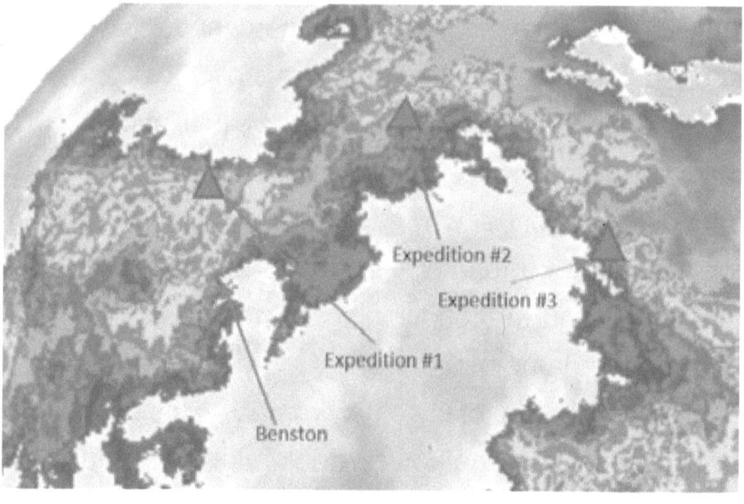

The Terran Sea Basin and expedition locations

CHAPTER 108

Expedition Number One

FOR EXPEDITION NUMBER One, the grasslands several hundred kilometers north of the colony were chosen because of the abundance of large migrating mammalian creatures. They speculated that hominids would locate themselves where food was abundant. The food source was discovered based on long-range drone flights. In fact, the southern end of the migration trail ended not far from Benston.

On the day the remaining *Calypso* shuttle craft lifted off from the field north of the colony walls, the monsoon season began in Benston. The warm air from the Terran Gulf, or the Terran Sea, as it was interchangeably called, pushed moisture north and for a solid two to three months it rained for a period of time just about every day. The temperature spiked, and the humidity made the colonists wonder why this location had been chosen for a colony. Fortunately, the electrical grid was fully functional, and air-conditioning helped push everyone through the damp months.

Terra and some of the other women on the expedition were seated on the lower deck along with all the equipment and stores needed for the two-month trip.

Hatcher was in the pilot seat on the upper deck. While in flight, Hatcher commanded the expedition. Once they were on the ground again, Gunner was in charge, but everyone knew that all the key decisions were subject to Terra's yay or nay, in the air or on the ground.

It would take about four hours of flying time to cover the distance of 1,249 kilometers. This could only be accomplished by flying at the shuttle's top speed when it was operating strictly in spinner mode. The route of travel was almost due north from Benston, and the expectation was that they would set down just inland from one of the great oceans. This ocean was called the Atlantic Ocean after the ocean on Earth. The climate on this coast was thought to be drier than Benston and colder by almost ten degrees centigrade, but the area was rich in animal life and had more than ample freshwater sources. For the first hour, the flight was bumpy as the warm air from the Terran Sea pushed over the colder air from the Atlantic Ocean.

Three hours into the flight, Hatcher's instrument panel displayed a red warning light. The light indicated heating in the rotational coolant unit. This was cause for concern as there was only enough fuel in the rocket propellant tanks for a single controlled descent, and no deceleration parachute had been packed as it was deemed to be unnecessary.

"I'm going to need everyone to buckle up. We have a warning light indicating overheating of the rotational unit. I need to look at it when we touch down. I'm looking for a suitable place to land. Hopefully, we'll be on the ground in about ten minutes," Hatcher said over the intercom.

There was a sudden jolt that threw everyone who was standing to the floor. The giant rotational blades had suddenly jammed and stopped in less than two seconds.

"Brace for impact!" Hatcher said over the intercom.

Terra and those below looked at one another in fear as they scrambled to buckle in.

On the flight deck, Gunner joined Hatcher in the copilot seat. "I can't feather the blades—they're jammed. I've got to try for a controlled descent with the propellant reserves. Help me look for some flat ground, Gunner," Hatcher said.

"Copy that," Gunner said. "All I see is forest, but I'll keep looking."

Hatcher fired the descent thrusters, which slowed their fall. But the altitude that he fired them at meant that they would run dry before they reached the ground, and that meant a crash landing. Each of their seats had pyrotechnic shock absorbers, but this was only good up to a certain g-force. Hatcher was almost certain they would exceed that.

"I've got eyes on a clearing, northwest of our position. I've got the cursor on it, about four kilometers at our ten o'clock position," Gunner said.

"I've got limited flight control functionality. I think our hydraulics have failed. I'm going to need your help to pull on the flight controls," Hatcher said as the descent engine ran dry, letting the shuttle begin another free fall. "Trying to release the blade gears again. Damn it! Wait, the blades are feathering. Prepare for impact."

The ground rushed up, and Hatcher deployed the landing legs seconds before impact. A gigantic jolt shook the entire shuttle as it made contact and then everything went dark and quiet.

CHAPTER 109

The Aftermath

TERRA WOKE WITH a very sore neck. So sore, she felt certain she had fractured a cervical vertebra. As she tried to stand, she immediately collapsed. Her fear was that she had a broken back and a damaged spinal cord, but she was able to stand on the second attempt. She looked around and saw others in the same condition. The smell of a burning metallic substance filled the lower bay. She looked around to see if there was a fire but didn't find one, just smoke.

Hatcher had just descended to the lower level and came up behind Terra and said, "It's boron oxides and other combustion products from the pyrotechnic seat shock absorbers. It's the reason we are still alive. Since the twenty-first century, helicopters have had various devices installed to take up the high g loads from hard landings. The support collar in your flight jumpsuit has a pyrotechnic device designed to keep your neck from breaking as our chins are thrown downward to our chest. Your neck is sore, your lower back is sore, but you are alive, thanks to centuries of safety device development."

"What's the situation?" Terra asked.

"We've landed about three hundred kilometers from where we wanted to land. Well, we've crash-landed. All a pilot ever hopes for is an equal number of takeoffs and landings," Hatcher said with his trademark gallows humor that instantly reassured Terra that at least *he* was OK.

"Casualties?" Terra asked.

"Gunner is in a lot of pain. He hadn't fully strapped himself in when he came up to help me. Some others from down below have some severe lower back pain. They could have fractured lumbar vertebrae. Nobody died, at least we have that going for us," Hatcher said with another forced smile. "And yes, my back hurts also, but it is bearable. Right now, we need to see if we can get some electrical power back and radio Benston to let them know what happened. I can't reboot anything from upstairs. Maybe the problem is at the power distribution rack. Can you help me with troubleshooting this?"

"Sure, lead the way, sweetheart," Terra said.

On the lower deck, they walked over to a hatch on the floor that led to the mini thorium nuclear reactor that powered the giant shuttle. After entering the secure access code, the door opened. A ladder led to the floor of the reactor room. Against the circular wall, amidst several equipment racks, was the power distribution rack. Thick bundles of fiber optic cables came into the rack, and smaller but more numerous fiber optic cables came out on the other side.

"Well, I see the problem now. The main power bus coming in has been pulled free from the last cable clamp, which allowed the mating connector to become unmated. The shock of the crash landing must have caused this. Likely the connector is damaged now, but perhaps we can reinsert the pins and sockets by hand and make it work temporarily. If we can get back to Benston, the connector can be repaired or replaced. I'm going to need that manual on the wall for schematics of the cabling, and I'm going to need more light in here. Can you round up some portable battery lighting for me?" Hatcher asked.

Ninety minutes later, Hatcher had patched the two sides of the fiber optic cables back together and secured them with gorilla tape. A check of continuity across the two sides showed that the repair was good.

"OK, now the moment of truth. The reactor did an emergency shutdown when we went down. I'm going to do a restart now. I'm really hoping nothing was damaged in the reactor due to the crash. Here goes," Hatcher said.

Hatcher and Terra waited fifteen minutes for the system to go through the self-checks that were part of a cold restart sequence of the reactor. When the last check finished, the reactor control panel showed that a successful restart had been accomplished. With power restored, lights around the shuttle came back on. They heard a smattering of clapping and cheering from above their heads.

"On to the next problem. I'm going to need to go outside and climb up to the top of the shuttle and check the spinner gearbox and see why it got jammed. That could be much more problematic. Is there anybody who can go up top with me?" Hatcher asked.

"I can," Terra answered, slightly annoyed at Hatcher for assuming she couldn't do it.

"It's dangerous, Terra," Hatcher said with concern for her.

"Hatcher, we traveled fifty light-years through interstellar space to settle on an alien world. Everything we do every day is dangerous. I'll help you," Terra said, softening her tone as she realized Hatcher loved her and was only concerned for her safety.

They exited out of the shuttle via the top emergency hatch. Hatcher had climbing harnesses that he attached each time he moved from one attachment point to the next. Terra followed him, carrying a backpack that contained the tools they needed. Several minutes of climbing and slithering up the domed surface ended with the two of them just underneath the giant rotor blades. An access panel covered the gearbox.

"Well, lookie here, the problem jumps out and bites us on the nose," Hatcher said as he removed the panel and shone his light inside the gearbox compartment. "Looks like the gearbox was recently opened up for preventative maintenance. Part of the disassembly steps is to install a holding pin that holds the alignment of the gear shafts when the gears are removed so that nondestructive inspections can be performed on them.

"Part of the pin is still inserted, but the pin broke when the gears started moving when we powered up the rotor for liftoff at Benston. The loose end of the pin bounced around inside as we flew. It eventually got jammed between the two biggest gears, which caused the failure of the rotors. It must have gotten worn down and eventually fallen out from between the two gears. That is why I was able to feather the blades and why we lived. We need to inspect all these gears. This is going to require that I remove all the gears on the shaft to get at the two that jammed. This will take some time. We're going to need more help to do this. These gears are big and heavy, and it will require a ton of cussing to get all this taken apart," he said with a smile and a sigh as he wiped sweat off his forehead.

The rotor blade gearbox disassembly and reassembly took two full days to complete. When completed, Hatcher was able to start the spinner assembly. The shuttle had sustained some other minor damage, which they began to repair. One of the four landing legs was bent, and it would not retract. Hatcher knew this was not optimum, but they could fly with the legs in the down position. The additional repairs would take another day or more.

The crew was still sore from their hard landing, but many of them wanted to get out of the shuttle and look around a bit. Gunner, who was confined to a gurney in the medical bay, advised against letting anyone out of the shuttle, but he was talked into it when they set a limit on the number outside at any time. It looked like they still had a couple of days to go before they could resume their journey.

CHAPTER 110

A Plea for Help

ON THE FOURTH day, at dawn, Terra took a walk outside with armed guards. They had landed on top of a mesa. The vegetation was coniferous trees, which covered the landscape except for where they had landed. On this mesa, for reasons unknown, the trees that grew everywhere else did not grow.

I guess that was fortunate for us, Terra thought.

She imagined that a crash landing among the trees would have left the shuttle damaged beyond repair. As she stopped to take in the view of distant mountains, she saw a glint of light through the trees near the bottom of the mesa.

"Jenny, can you see a light among the trees at the bottom of the mesa?" Terra asked Jenny Whitehorse who was walking with her.

"Where exactly?" Jenny asked.

"Down there." Terra pointed to the bottom of the hillside.

"OK yes, it looks like a campfire."

As they looked into the trees at the bottom, they saw a handful of people emerging from the tree line. They both knew that they had been seen by the people at the bottom of the hill. Their initial reaction was to drop to the ground and make themselves not visible, but they realized it was too late. They had been seen, and these people, whoever they were, were now waving their arms to get their attention.

Terra spoke into her comm device. "Hatcher, we've got company, bottom of the hill, north side. We've been seen. I've got a dozen or more people moving up the hill waving their arms at us. I'm going to try to read them, get a feel for their intentions."

"Copy that, Terra," Hatcher responded.

As Terra stood back up, she raised her arms to get the attention of the people at the bottom. They waved back and began moving more quickly up the hill to reach Terra and Jenny. Terra could sense fear, desperation, and some hope among the people as they advanced.

"Hatcher, I think these people are fleeing some sort of danger. I don't think they are dangerous. They need help," Terra said into her comm device.

"Copy that. I'm coming out now with Joshua and Thomas. Stay where you are," Hatcher said.

The group reached the top of the hill and walked over to Terra and Jenny. Hatcher, Joshua, and Thomas arrived at the same time along with the guards who were outside with Terra. A man, bearded and very thin, spoke in his native tongue. Terra turned to Joshua and Thomas to see if they understood the man. They both said they understood some of what he said, but the dialect was not the one in which they were fluent. They said the man explained that they were running from some type of danger and many of their people were dead. He also said they were hungry.

Terra looked at the man closely; then she walked over to him and hugged him, holding him for over a minute. She released the man and then turned to Hatcher and the others and said, "Their people were attacked by hordes of people on horseback who arrived by boats. Some of their people escaped—most were killed. Those who escaped were being hunted down and killed by the people on horseback. The people hunting them are only a few hours behind them. They want us to take them in and help them. They also said they fear for others who will eventually be attacked by these people. They want to warn us to leave and go far away to avoid the same fate their people suffered."

CHAPTER 111

Decisions

THEY BROUGHT THE group inside the shuttle and gave them something to eat. Meanwhile, Hatcher, Terra, Gunner, Joshua, and Thomas huddled around Gunner's gurney to discuss what to do.

"Joshua, Thomas, do either of you know anything about these people with boats and horses?" Terra asked.

"Many moons ago, before we were born, our people and the other coastal clans were attacked, and many were killed by people such as these. Perhaps it is the same people. If so, the thing you should know is that they are very cruel. They killed men, women, and children. They have weapons made of metal. What you call them? Long knives?" Joshua asked.

"Probably swords and other metal armament like shields, axes, and such," Gunner added.

"What did they want beyond killing?" Hatcher asked.

"They wanted everything of ours, our furs, our food, but more than anything else, they were looking for certain types of women," Thomas said.

Hatcher looked at Terra with an unspoken realization between the two of them.

"What type of women?" Gunner asked.

"Women like Terra," Thomas answered.

They all stopped talking and sat with the realization that they somehow knew about The Mothers.

"Well," Gunner said. "I am sure we can handle them."

"It is said that they came in very large numbers, like the grains of sand on a beach," Joshua said soberly. "They may have even better weapons now."

"If they have ships and horses, they may be able to travel great distances with great speed. We need to get eyes on these people. Figure out where they are and where they are going. We need to warn Benston about this threat and prepare accordingly. Terra, can you see if these people can tell us where those people are or at least where they were?" Gunner asked.

"I can try," Terra said. "It's easy enough for me to know what they are thinking, but if I try to enter their minds to ask questions, they won't understand our language. I can try to draw pictures that help them understand what we want to know."

"Perhaps what would be best is to capture one of these people and have Terra look into their minds, figure out where they came from, where they are going, and why they are here," Hatcher said.

"I agree. In the meantime, let's put up some drones to get an idea of what's out there. If we happen to locate them, we can plan to capture one of them, but it is going to be very risky. We need to think this through very carefully before we execute a plan like that," Gunner said, wincing with pain as he finished. "Can somebody get Jenny and ask her if it is time for another pain shot?"

Terra answered, "Sure, I'll find her."

* * *

"How are you feeling?" Jenny asked Gunner.

"I've got pain down my right leg and down both arms," Gunner answered.

"Sciatic pain. You have a cracked vertebrae at L5 and two cracked vertebrae in your cervical spine. Looks like C4 and C5. Our portable X-ray cannot tell me more than that. When we get back, you are looking at surgery to stabilize those locations while they heal," Jenny explained. "I am so sorry, I know you are in pain. The important thing is to stay immobilized. I know you do not like the catheter and the bedpan, but they are necessary so that you don't damage your spinal cord."

She could see that Gunner was frustrated and in a lot of pain. She held his hand and squeezed it to let him know she cared. Gunner smiled and squeezed back.

Meanwhile, Terra was sitting with two of the refugees. She had a pencil and paper and had drawn the mesa they were on, the shuttle, and the mountains to the north. She also drew a sunrise, to indicate the direction of east. She drew a horse and a man on the horse and then tried to convey that she wanted to know where the men on horses were. It seemed to work as the skinny older man nodded his head and pointed to the horse and man and then pointed north of where the shuttle was located on the mesa. Terra smiled and hugged the man, all the while reading his thoughts, which confirmed that he had understood what she wanted to know.

A little while later, Hatcher launched two drones northward, one traveling a greater distance, one closer, and had them search in a pattern that ended up with both meeting in the middle distance. The automatic search mode was given parameters looking for horses and men. A third drone circled the mesa, checking for anything coming up the hillside. Hatcher had three sets of guards stationed around the hilltop looking for anything coming up the hillside or anything that could be seen in the dense forest below. About ninety minutes later, the more distant drone began pinging an alarm, indicating it had found items matching its search profile.

Hatcher, Joshua, and Thomas gathered around Gunner's gurney and watched the live feed from the drone. What they saw surprised each of them. A group of horses and men were stopped in a clearing next to a river. Several men had dismounted and were talking to one another. Slung across the backs of each man were some sort of rifles. As they watched, one of the men motioned for the others to be quiet. They stood there looking into the sky for a moment.

Hatcher manually commanded the drone to mark the location and ascend into the clouds. He switched to infrared mode, and the same view appeared again with the men and horses looking like warm bodies, no longer clear daylight types of images. The men got back onto their horses and rode back north and east. Eventually, after an hour of riding, they entered a much larger clearing that was filled with men, horses, and tents. This appeared to be the main camp, and it was impossible to count the number of men as there were probably four or five thousand. What surprised Hatcher and Gunner even more was that there were cannons hooked up to teams of horses.

"Horses, rifles, cannons, and thousands of men. These people are serious about what they are up to. We need to capture one of these people," Gunner said.

CHAPTER 112

Capture

THE CAPTURE PLAN was developed after three days of watching the main encampment and the scouting sorties that were sent out. The invaders seemed to have camped for a while in a semipermanent location. The Earth people had observed a steady growth of the encampment, and each day, more men on horseback arrived from the north. Gunner had asked for a long-range drone to go farther north in hopes of finding where they had arrived, presumably by ship.

The long-range drone was finally in place and searching the coast for ships. The process was slow as the drone could only receive communication with the shuttle when it was airborne at an altitude where comms would work across the landscape between the shuttle and drone. Each day, the drone would search for three hours then listen for new instructions; then it would land and let the Terran sun recharge its batteries. After searching for a week up and down the northern coast, a vast number of sailing ships were located in a secluded series of harbors. Whoever these people were, they knew how to build ships and navigate the great bodies of water on Terra Nova.

“I think we want to try to draw the patrols they send out toward our location. There’s never more than a dozen of them. If we have them chase one of our people, we can lead them into an ambush and hit them with electrostatic pulses from our stun guns. This will stun the horses as well. We’ll come in, blindfold them before they wake up, and restrain them. We will take their horses with us and then take a couple of the riders back with us. We don’t want them to see what we have up here on the mesa, so we’ll bring Terra down and meet her at a predetermined location. She can read them and then we blindfold them again and take them back away from this mesa and release them,” Hatcher said.

“Why release them? If we keep them, we could learn more about them. We could bring them back with us and question them more back at Benston. We need to know how these people got more advanced weapons than any other group of people here. There must be more of a story behind all that,” Gunner said.

“I agree with Gunner. Maybe there were AOW spies here among The Mother’s people before we got here. We need to know as much as we can because we will need to redirect these people away from their violent tendencies,” Terra said.

“OK, let’s do that then. I think we want Joshua or Thomas to be the bait. If they see us, they may get spooked and not pursue them into the ambush. They look like the people they expect to see. Are you guys up to this?” Hatcher asked.

Joshua and Thomas both smiled and nodded.

"Good, let's get a large enough group that is well armed and head down the hill and find a good ambush spot," Hatcher said.

There were hills and streams in this part of the forest, and this ended up providing the ambush Hatcher wanted. A twenty-meter waterfall that dropped into the lower elevations of the forest provided the perfect spot for Joshua and Thomas to run to and a spot to trap the pursuers. The drop-off was too steep for horses, and this would hinder the likelihood of the riders simply fleeing the field from this direction. Boulders would be placed on a line perpendicular to the ledge on both sides of the stream that would slow the riders down and dissuade them from fleeing along the rim. The rim was devoid of trees, and it would have been easy for horses to gallop away. Eight Earthers on each of the three sides would be hidden as the two Terrans ran to the waterfall. Stun guns set on high would ensure that the horses and men would be rendered unconscious or seriously stunned. Meanwhile, two captives would be secured, blindfolded, and taken back to the shuttle. The remaining riders would be bound and blindfolded.

Hatcher had the group practice the ambush twice before they attempted the real thing. The opportunity to do the ambush would have to wait for the patrol to enter this area of the forest. A drone was watching the large camp, so it was possible to see a patrol leave and head off in the direction of the shuttle. The ride was several hours from the camp to the stream, which was north of the mesa where the shuttle had crashed. The main group at the shuttle would climb off the mesa when the drone confirmed that the riders were headed in their direction and would take up their positions.

Two days after the last practice ambush, eleven days after the crash, the drone over the encampment picked up a likely patrol that left in the direction of the shuttle. Hatcher led the ambush team down off the mesa. Within ninety minutes, they were in place, hidden near the stream and ledge. Hatcher was with Joshua and Thomas on what they believed was a trail in the forest, hopefully, the one the riders would follow. From where they were, the waterfall and the hidden Earthers were about three hundred meters distant. Hatcher had Dale LaPlant in command back at the waterfall. The two were in radio contact as the riders did indeed choose to follow the trail Hatcher and the Terrans were hiding alongside.

Per the plan, Hatcher left Joshua and Thomas with radios and started to head off to the waterfall. Hatcher had told the two to speak the words "broken arrow" if they were in danger and would not make it to the waterfall. Hatcher had each of the two drones equipped with stun grenades to slow the riders down if necessary. Two additional Earthers were hidden further up the trail from where Hatcher and the Terrans were hidden. They were to let Hatcher know when the riders got close, and they were to slow the riders down if Joshua and Thomas were captured, allowing the remaining Earthers to catch up and help them free the Terrans, if necessary. In reality, Hatcher didn't know if they would bother capturing the two Terrans or maybe they would just kill them. Hatcher tried not to think about this. They were close friends of his after their experience with the sage cat.

As that less-than-cheery thought slipped from his processor's register, thundering hooves could be heard coming down the trail. Joshua and Thomas had stood up as Hatcher entered the forest to slip away and join the others at the ambush location.

The riders seemed to appear out of nowhere. There had been no notification from Elias and Jed, the two Earthers up the trail who were supposed to warn them of the imminent arrival of the riders. Joshua and Thomas were caught out in the open on the trail and started to run down the trail per the plan. The crack of a rifle sounded, and Joshua dropped to the ground, having taken a bullet to his leg. Thomas stopped to help Joshua, and he was hit by a bullet that killed him instantly. Hatcher screamed and ran toward the riders, his stun gun in his hands. As he got close enough to use the weapon, he saw that the riders had Elias and Jed tied and bound on a horse. One of the riders dismounted and took a knife out and slit the throat of Jed Diamond, a geologist who had just married his longtime girlfriend.

The dismounted rider held up the bloody knife and shouted to Hatcher in perfect English, "Drop the stun gun or I slit the other one's throat as well!"

Hatcher froze, raised his gun for a moment, then tossed it to the ground. The dismounted man told him to come out of the forest and onto the trail. Hatcher faked a misstep, reached for his radio, and said softly, "Broken arrow, broken arrow."

The rider shouted at Hatcher, "You said something over the radio! What did you say?"

As he said this, he went over to Elias and held the knife to his throat. "Last chance, what did you say on the radio?"

Hatcher hesitated for a second then cleared his throat as he got up and said, "Broken arrow. I . . . uh, said broken arrow."

"And what does broken arrow mean and who were you talking to?" the rider asked.

Hatcher answered, "It means we've been captured, and our soldiers will be coming to rescue us and kill you all!"

A rider who had dismounted walked over to Thomas's body where Hatcher was and swung his rifle at the back of Hatcher's head, and everything went dark.

CHAPTER 113

Chaos and Clarity of Purpose

THE NEXT TWO hours were chaotic and confusing. Dale LaPlant radioed back to the shuttle, explaining that Hatcher and the two Terrans were likely captured and that the status of Elias and Jed was unknown. At the shuttle, Gunner assumed control of the situation.

"I want that long-range drone on the coast to head to the encampment and watch for the return of the patrol. We need to have eyes on what is happening there, but we also need eyes on that trail they were on," Gunner said. "We need to have the trail drone hover over the point where Hatcher and the Terrans were last known to be. Until we know the situation with Hatcher, Elias, Jed, and the Terrans, I do not want to expose any more of our people."

"Hatcher specifically had the drones loaded with stun grenades. If we don't slow down the riders, we may not get a chance to get our people back," Peter Chan said. Peter had been a late addition to the shuttle personnel after Walt Stephenson developed a kidney stone and was scratched from the roster.

"I hear you, but instead of us ambushing the primitive natives, they ambushed us. I'm done underestimating who these people are. We will get our people back, but it will be my way. Terra, can you sense anything from Hatcher or the Terrans?" Gunner asked.

"I can't sense anything from Hatcher or Thomas," Terra said with a shaky voice. "I don't know what that means."

"What about Joshua?" Gunner asked.

"Let me focus on him," Terra answered. "Give me a few minutes."

Terra went downstairs and asked not to be disturbed as she opened the door to the cargo locker. Terra slumped to the floor and began sobbing. After a minute, she started singing The Mother's song.

As she sang, Eve appeared. "Eve, I need your help. I told you yesterday about the riders. Today, we had planned to capture one of them to determine their intentions. It didn't work. Hatcher and the others were captured. What should we do now?" Terra broke into sobs again.

"You must NOT go after these men. They are very dark souls. What they seek is you. They will use Hatcher, Joshua, and Elias to force you to come with them. You and your people will return to Benston and await their ultimatum. This group of Terrans was trained by AOW spies long ago. The spies have all died, but they gave technology to and trained the Clan of the Horsemen in warfare. The Clan of the Horsemen seeks dominion over all the clans in this world. They must be defeated, but this will not be through bloodshed. Defeat will come to them in our unique way. Their hearts must be changed. At the appropriate time, I will tell you how this will be done," Eve finished and then left Terra alone, wondering what Eve meant.

Terra went back to the upper deck where Gunner's gurney had been moved and spoke to the assembled group. "I spoke with Eve. We are to return home and await their ultimatum. They will be dealt with, but not by force of arms."

Later that day, the bodies of Thomas and Jed were picked up and returned to the shuttle. At dawn on the next day, Gunner was helped into the pilot seat with Peter Chan in the copilot seat to help him. After preflight checks were complete, the shuttle lifted off and headed back to Benston.

CHAPTER 114

Prisoners

HATCHER WOKE UP on the ride back to the encampment. His head hurt viciously, causing him to become nauseous, resulting in him vomiting on the side of the horse to which he had been slung over and tied.

"Can I be set up on the horse? You can leave my hands tied. I won't have to vomit on your horse again if you do this, I promise," Hatcher said, hoping a little humor would help their situation out.

One of the riders said something in a tongue Hatcher did not understand to another rider, who got off and cut Hatcher's restraints, allowing him to sit up without even his hands tied.

"If you try to escape, we will kill your wounded friend. If you do escape, we will catch you and torture you in the most unpleasant manner and then we will leave you to die slowly. Cooperate and you will live to be traded back to your people," the lead rider said, this time in perfect English.

"What do you want from my people?" Hatcher asked, genuinely curious.

"Your Oracle, the woman named Terra."

Hatcher was stunned that they knew about Terra. He considered what to say in response and decided the less they knew, the better. He said nothing.

A little while later, Elias regained consciousness. He had been knocked unconscious when he and Jed were captured. His hands were freed, and he was set on his horse like Hatcher.

As for Joshua, his leg wound had bled a lot, but the riders had put some bandages on the location where the bullet had entered his upper leg. The bleeding had all but stopped, which Hatcher thought was a good sign. When he regained consciousness, he was placed upright on his horse as well. Joshua did not look well. He was pale and seemed to have trouble staying on the horse. Hatcher was taken off his horse and placed on Joshua's horse so that he could hold him in the saddle.

Joshua whispered to Hatcher, "Is Thomas dead?"

"Yes, he is. I'm so sorry," Hatcher whispered.

Joshua remained silent for a moment then whispered, "I will kill their leader to avenge Thomas. I will cut his head off and mount it on a pole."

"Right now, don't try anything. They said something about releasing us to our people through a trade. Let's see how that works out. Revenge can come later," Hatcher said in a hushed tone.

After three and a half hours, they arrived at a giant clearing full of tents that seemed to stretch to the horizon. They were taken to one of the tents, and a woman sewed up the entry and exit wounds on Joshua's thigh.

Afterward, the same woman put a salve on Joshua's leg and rebandaged it.

They were then taken to a pit covered by a tarp and put into it. The walls were smooth, and the pit was well over five meters deep. There was not an obvious way out. Two guards remained at the top of the pit just in case they did find a way to reach ground level.

On the floor of the pit were straw and three blankets. The weather had turned cold in the evening, with a light rain falling. The three prisoners huddled together trying to stay warm.

"I think you two were still out when they told me what they wanted to trade us for," Hatcher said. "They want Terra. I think we might be here for a while."

"How's your leg?" Elias asked Joshua.

"It hurts, but I can walk."

"Good, because we need to start observing the behavior of the guards, see where and when we can make a run for it," Elias said.

"I think we could make a run for it. It might take us a couple of months, but we could make it back to Benston. I've been hunting the game here for a while now. I think we could survive in the wild, living off the land as it were," Elias added.

He had been a hunting guide and a naturalist on Earth. If anyone could pull off an overland trek of a thousand kilometers, it was him.

"Joshua might need a week or two to get healed up. His leg could get infected. I think the ointment they put on your leg is like pine resin. It has antiseptic properties. These people have been given a real leg up on the development ladder," Hatcher added.

"I agree," Elias said. "Let's keep a close watch on the comings and goings here for at least a few days before we try anything."

"I should tell you that they promised they would kill us if we tried to escape. In fact, they promised a slow and painful death. So whatever we do, it must work," Hatcher added soberly.

On that less than cheerful note, they huddled together on the straw and tried to sleep, which came surprisingly easy, given the eventful day they had.

CHAPTER 115

Watchful Waiting

"HAND ME THE bone glue and some pads. This is worse than I first suspected. I've got six lumbar fractures and three cervical fractures," Dr. Holly Guzman said with no emotion at all.

Dr. Sampson, who was assisting Dr. Guzman, a neurosurgeon, asked, "No hardware needed then?"

"Nope, just bone glue with his own stem cells and one week of the acceleration treatment. He'll be back in top form in six weeks. Thank God we live in the twenty-fourth century. Even just three hundred years ago, Mr. Gunner Lawson would have ended up with nine fusions, a bucket full of ghastly metal hardware, an early retirement, and most tragically, an opioid addiction and associated organ damage. Aren't we all so lucky!" Holly said with no empathy, clearly exposing her atrocious bedside manner.

Fortunately for her, she was one of the best spinal repair surgeons on Earth, and that had forced everyone around her to ignore her sociopathy. Now that the Earth was gone, Holly was certain she was the best remaining surgeon in any specialty, and that was indeed a powerful drug for her.

Gunner's surgery and a weeklong stay in the bone-growth stimulator tube had left him completely out of the loop on any plans to get Hatcher, Elias, and Joshua back. He had heard about what Eve had told Terra to do, but he still wasn't accepting that waiting was the best approach. This had to be dealt with differently.

* * *

Gunner left Simi, Taylor, and their kids in his house. He needed to get outside and think through a strategy. As he walked to the settlement gate and exited it, he almost ran into Terra who was walking slowly and seemed pensive.

"Terra, what's going on with Hatcher and the others? I haven't heard anything since we got back," Gunner asked.

"Hatcher, Elias, and Joshua are in a pit in the horsemen's encampment," said Terra. "They don't know what is happening. The riders are not sure about what to do. They realize that we are working with The Mothers, and they know about me. They are trying to determine if a trade can be made."

"What trade?" Gunner asked.

"Their khan wants me. He knows what I can do, and he wants me," she said.

"Well, we can't do that. We're going to have to go get them somehow."

"How would you propose to do that? Our drone has shown that their encampment now has over ten thousand men and hundreds of horses. They have guns and even cannons. They could surround us, starve us out, and get what they want, or we could get back our people, and I can be given to the khan. I believe they would leave us be if we did this," she said.

"What is a 'khan' anyway?" Gunner asked.

"The word *khan* means leader. It has its origin from Earth history. You recall Genghis Khan?"

"Yes, but how would they know about Earth history?" he asked.

"Some of The Mother's people who lived here before we arrived were AOW spies. They must have taught them about Earth history and gave them gun technology. I was thinking there might be a way to communicate with the riders. Hatcher, Thomas, and Elias all had handheld radios. I know their range is mostly line of sight, except at nighttime with cloud cover, but we could put a line of drones up at high altitudes and speak to them by using the drones as comm relays. Of course, they must know something about radios, but it is worth a try," Terra said.

"We should try that. Have you talked to President Davis about this?"

"Not yet. Do you want to come with me? He's usually in his office at this time of day," Terra said.

"Sure," he answered. "Let's go talk to him."

* * *

"We can spread the drones about four hundred kilometers apart at an altitude of about three thousand seven hundred meters. That should give us a chance of reaching their encampment, assuming someone is listening and willing to answer the phone," Peter said. "It will take a couple of days to get the last drone in place. The travel time, you know."

President Davis said, "OK, make it happen, Peter. Gunner, how are the repairs on the shuttle going?"

"Well, I haven't been involved. I've been in a coma in an acceleration tube. It was healing my back and my vertebrae. But I'll find out and let you know," Gunner said.

At the end of the week, on Saturday, the last drone had reached its necessary location and had landed and charged itself up. By 9:00 a.m., the drone was loitering at three thousand seven hundred meters, about twelve thousand feet altitude. In the comm room of the shuttle, Terra, Taylor, Gunner, Simi, Peter, and President Davis sat around the comm console.

"Terra, if you talk to these people, as I understand it, you can get a feel for their intentions. Is that correct?" President Davis asked.

"Yes, as soon as someone directs their thoughts to me, I can hold on to their thoughts. Let me talk to them, Mr. President," Terra said.

"OK, go ahead," he answered.

"Hello, is anyone receiving me? Over," Terra asked into the comm speaker.

Instantly, a gaggle of voices could be heard on the other end. Their language was unfamiliar.

"I want to speak to your prisoners. Get me Hatcher. Let me talk to him, over," Terra said.

More voices, speaking quickly on the other end.

"Wait." Finally, a word in English from the other end.

"Do you need to get them?" Terra said. "I can wait, over."

"Put the speaker on mute for a moment," Gunner said. "Can you read anything from their thoughts?"

"Yes, they were expecting us to call," Terra answered.

CHAPTER 116

The Longest Day

"TERRA, IS THAT you? Can you hear me?" Hatcher asked, out of breath. It had been about ten minutes of waiting before he came on the radio.

"How are you doing? How are the others? Joshua and Elias?" Terra asked excitedly.

"We're OK. Joshua got shot in the leg, but he seems to be healing OK," Hatcher said, then he was cut off.

"I am the Khan. I am Ammon Khan, leader of the people of horses and ships. We want to speak to the woman named Terra," the Khan said.

"I am Terra."

"We want no war with your people. But you have something that we want, and we have something that you want," he said.

"I know what you want. You want me, and we want our people back."

"Yes, so you are in fact an Oracle," the khan said.

"Yes, I am. Eve of Eden, our great Mother, anointed me The Mother of all Terran Peoples," Terra said.

"So you are. Then you know you must become my wife and mother of my children. We are to rule together, the great Khan and the great Mother Oracle," he said, with no lack of confidence or firmness.

"I am married to another already. I am a mother of two small children. I can be your Oracle, but I cannot be your wife," she said.

"Then your people here are of no value to us and will be killed very slowly, and we will come to your settlement and surround your walls and starve your people. You must submit and become my wife! You have until tomorrow at this time to agree, or you will all die!" the Khan thundered.

"Sounds like somebody needs anger management gene splicing!" Gunner said.

"Would we even know how to do it?" President Davis asked.

"I know those genes were identified and removed from the human genome over a hundred years ago, and now no one has the gene combinations," Peter said. "I suppose we could rediscover which ones were removed and try it on him."

"You know I was just trying to lighten the mood. But here you two geeks have swallowed my hook, line, and sinker!" Gunner said with a laugh.

Simi poked Gunner in the side, indicating her disapproval toward his attempt at humor.

"Simi and Taylor, can you get my parents and meet me back at my house? President Davis, I need some time to think this over," Terra said.

"Wait, nobody is taking this seriously, Terra. We will not give you to this barbarian!" President Davis said. "We need to plan an assault on their encampment! Gunner, get the Congress assembled. We need to make our plan of action."

"President Davis, there is no action to be taken. We would never achieve our goals here on this planet if we were to attack their camp and kill thousands of their people. We would still lose our people. We will not proceed that way. Give me some time. Everyone, go about your day. We will survive this. Please don't trouble yourselves," Terra said as she walked out with Taylor and Simi.

Terra, Taylor, Simi, Terra's parents, and Hatcher's parents met up in Terra's house.

"I have a very difficult choice to make. The leader of the horse people, Ammon Khan, is threatening to kill Hatcher, Elias, and Joshua unless I agree to his terms. He wants to take me as his wife. I can't see a way around this problem. We have until tomorrow to agree to this, or he will kill our people. If I agree to his demands, perhaps I will be able to influence the Khan and his people to adopt more peaceful ways. But this would mean that my marriage to Hatcher would be over, at least for a time. I don't know how old this man is. I will not age, Hatcher was promised a very long life, perhaps we can be together again," Terra said as she broke down into tears, shaking and sobbing in the arms of Taylor and her mother.

Nathan, Hatcher's father, spoke up. "We could threaten to destroy their camp. The shuttle has a rail gun. That would cause great losses among their people."

"Nathan, Sherri, I am afraid that they would still kill Hatcher and the others. I can't think of a way where we use force that doesn't result in us losing our people. If we kill their people, they will never trust us and it will start a cycle of violence," Terra said. "Think of how the cycle of violence led to the nuclear war and the deaths of all the Jewish and Arabic peoples. We can't let that type of violence begin here."

"What will happen to your twins, Terra?" Riley asked.

"I cannot put them into an uncertain and dangerous environment. They will stay here."

"I will care for them," Taylor said.

"What will we tell them?" Simi asked.

Sam said, "We will tell them the truth."

"How will they ever accept that?" Riley asked between tearful pauses.

"They will come to accept it," Sam said. "They are young."

CHAPTER 117

Mother of the Terran Peoples

THE NEXT MORNING, Terra got up after a sleepless night. The sun was still not up, but the eastern horizon was turning a glorious yellow. The walk to the beach that she and Hatcher used to walk took almost thirty minutes. As she walked to the water's edge, the sun began to rise above the horizon.

Her thoughts raced, and her emotions overcame her as she started crying softly. She had come so far in her young life, but she had always had an idea of what lay ahead. Now she had none. She began humming, as this had almost always made her happy. She began to hum The Mother's song. She needed Eve. She sang through tears:

Ancient Mother, I hear you calling me,
Ancient Mother, I hear your song,
Ancient Mother, I hear your laughter,
Ancient Mother, I taste your tears.
You are the water that I drink,
You are the air that I breathe,
You are the sun who gives me warmth,

You are the earth beneath my feet.
Open your heart, open your mind,
Feel your soul, fly your dreams.
(Ancient Mother by Sacred Earth)

As she finished, she felt a presence behind her. She turned and saw Eve. She fell into her arms and cried, but now she had a feeling of clarity and certainty.

At that moment, she knew how she would become The Mother of the Terran peoples.

CHAPTER 118

Farewell

Benston Colony
September 1, 2354

THE SHUTTLECRAFT LIFTED slowly off the ground north of the colony walls. The giant blades were always a source of awe for the children of the colony. The wind they produced to lift such a giant craft was impressive. So much so that a blare of the colony warning siren was sounded to warn colonists about possible dust and debris entering their homes.

Terra sat on the lower level of the shuttlecraft. Her face was stained with tears she had shed that morning as she bade farewell to her children, her parents, Taylor, and Hatcher's parents, as well as pretty much everyone else in the colony. Gunner was piloting the shuttle with Peter's help. A small contingent of armed colonists was on board to assist with any contingency. They were heading back to the mesa where they had crash-landed six weeks earlier. The khan and some of his men were going to meet them there at midday.

Gunner was still certain that they could overpower the riders and take back their people without giving up Terra. Terra was not at all certain of that. At any rate, the Khan now had twenty thousand men stationed somewhere within a few hours of the colony to threaten the colonists if they acted in bad faith in turning over Terra.

As they approached the mesa, Gunner had the shuttle circle the hilltop to assess what strength the riders had in or around the mesa. There was a large encampment on the hilltop, which was going to be in the way of a safe landing. Gunner decided that setting the shuttle down would force the riders to yield the hill to avoid being hit by the landing shuttle.

"Assholes are going to lose their tin-pot headgear," Gunner said with a sadistic smile.

As the shuttle hovered for descent and landing, the men on the hilltop did, in fact, seem to get frightened and moved off to the south side of the hill where a trail allowed access to the top of the mesa.

As the shuttle touched down, Gunner ejected a drone to circle the shuttle and assess threats. Satisfied that there would be no rush by the riders to attempt a takeover of the shuttle, Gunner ordered the armed colonists to leave the shuttle and take up defensive positions around the craft. As they did this, a line of horsemen moved closer to the shuttle. On three of the horses were three blindfolded men. Gunner assumed these were Hatcher, Elias, and Joshua. Gunner had Peter act as the liaison on the ground while he remained at the controls of the shuttle in case a quick dust-off was required.

"We have Terra as you demanded," Peter said to the three riders who approached the shuttle with the three blindfolded men.

The lead rider dismounted and walked toward Peter. He was a giant of a man, probably 213 centimeters tall, with massive metal armor on and a face shield that covered most of his face. He removed the helmet, and shoulder-length dark hair fell out of the helmet, revealing a surprisingly handsome bearded face. He strode directly up to Peter and offered his hand to him. Peter stood for a second or so, unsure of what to do. He reached out his hand and shook the offered hand.

"I am Ammon Khan, leader of the nomadic peoples of the great plains to the north and east of here," he said in perfect English. "May I see the Oracle?"

"Yes," Peter answered as he turned back to the shuttle entrance where Terra waited.

Terra stepped off the ramp and walked over to Peter. She was clothed in a green cloak with a hood. As she reached Peter, she pulled off the hood, and her long dark hair fell down her back.

"I am Terra, the one you refer to as the Oracle."

"Terra, I am pleased to make your acquaintance," Ammon Khan answered.

"We have your people here. One of them, Joshua, was injured by one of my people. I am sorry for the needless loss of life caused by my men. I cannot take back their actions, and I can only offer my apology. The wounded man may need the help of your healers. We have done what we could to help him recover, but I am told he may have an infection by one of these men, one of your sky people, Hatcher," the Khan said.

"Thank you for your attention to his injuries," Peter said.

"We are looking for a harbor for our ships and a place to winter. I am told that the seasons are gentle where your colony is located. Is there a harbor there that we may use to anchor our ships?" the Khan asked.

Peter said nothing for a moment, then Terra answered, "Yes, Ammon Khan, just south of our colony is a bay that is protected from the sea, with a large fertile plain where your people may rest from their travels and await favorable sea winds in the springtime. We would be most happy to let your people stay as long as they wish."

"Then we shall return to our ships and sail to that harbor. Do you have maps to show us the way? Or perhaps you could spare a man who can navigate us to the harbor?" the Khan asked.

"This man, Peter"—she pointed to Peter—"is our best navigator. He helped in making maps of the entirety of this world. If he is amenable, he could go with you and help guide your ships there," Terra said.

"How far of a distance is this?" the khan asked.

"I can only guess without the maps in front of me, but probably about two to three weeks, depending on the winds. We would need to sail south around this large peninsula, then north up to the harbor. The sea bounding our colony we call the Terran Gulf," Peter said.

In the end, Peter agreed to accompany the riders back to their camp and then go to their ships to help guide them to the Benston colony.

"I see you have named the great sea after my new wife, the Oracle. Aptly named. Peter, let me release your men to you. We can take Terra at this time," the Khan said.

"Ammon Khan, may I speak privately to these men you are returning to us? They are dear friends, and I will miss them," Terra said.

"Yes, please make your farewells," the khan said, motioning to the three hooded men.

Terra walked over to the three men who had now dismounted their horses and were no longer hooded.

She came to Hatcher and hugged him. She whispered into his ear, "It is best if he does not know that you are my husband. I could not think of a way to save your life without agreeing to his demand that I become his wife. I am so sorry, my love. Take care of our children," she said as she fought back tears while releasing Hatcher.

Hatcher watched her go. As she walked away, he dropped his head. He realized he had no choice but to let her leave.

Terra was helped onto a horse next to the Khan's mighty steed. The riders formed a single file line on their horses as they rode to the edge of the hillside trail leading off the mesa.

Gunner motioned for the colonists to reboard the shuttle. They all began moving to the ramp leading to the door. Hatcher was the last one to board.

As he walked up the ramp to the open door, he stopped at the top and took one last look at Terra.

Terra saw him looking and slowed her horse to a stop and waved one last goodbye, with tears running down her face.

The future was a complete blank scene to her. She had no idea what it held for her and Hatcher, though she was determined that this would not be the last chapter in their life together.

The End of Book 1

www.ingramcontent.com/pod-product-compliance
Lightning Source LLC
LaVergne TN
LVHW100132061126
851857LV00010B/64
9798901242902